Conference Proceedings of the Society for Experimental Mechanics Series

Series Editor

Kristin B. Zimmerman, Ph.D.
Society for Experimental Mechanics, Inc.,
Bethel, CT, USA

Conference Proceedings of the Society for Experimental Mechanics Series

Series Editor

Kristin B. Zimmerman, Ph.D.
Society for Experimental Mechanics, Inc.,
Bethel, CT, USA

The Conference Proceedings of the Society for Experimental Mechanics Series presents early findings and case studies from a wide range of fundamental and applied work across the broad range of fields that comprise Experimental Mechanics. Series volumes follow the principle tracks or focus topics featured in each of the Society's two annual conferences: IMAC, A Conference and Exposition on Structural Dynamics, and the Society's Annual Conference & Exposition and will address critical areas of interest to researchers and design engineers working in all areas of Structural Dynamics, Solid Mechanics and Materials Research.

More information about this series at http://www.springer.com/series/8922

Matthew S. Allen • Walter D'Ambrogio • Dan Roettgen
Editors

Dynamic Substructures, Volume 4

Proceedings of the 39th IMAC, A Conference and Exposition on Structural Dynamics 2021

Springer

Editors
Matthew S. Allen
Brigham Young University
Provo, UT, USA

Dan Roettgen
Sandia National Laboratories
Albuquerque, NM, USA

Walter D'Ambrogio
DIIIE
University of L'Aquila
L'Aquila, L'Aquila, Italy

ISSN 2191-5644 ISSN 2191-5652 (electronic)
Conference Proceedings of the Society for Experimental Mechanics Series
ISBN 978-3-030-75909-4 ISBN 978-3-030-75910-0 (eBook)
https://doi.org/10.1007/978-3-030-75910-0

This Springer imprint is published by the registered company Springer Nature Switzerland AG
The registered company address is: Gewerbestrasse 11, 6330 Cham, Switzerland

Preface

Dynamic Substructures represents one of nine volumes of technical papers presented at the thirty-ninth IMAC, A Conference and Exposition on Structural Dynamics, organized by the Society for Experimental Mechanics, and held February 8–11, 2021. The full proceedings also include volumes on nonlinear structures and systems; dynamics of civil structures; model validation and uncertainty quantification; special topics in structural dynamics and experimental techniques; rotating machinery, optical methods, and scanning ldv methods; sensors and instrumentation, aircraft/aerospace, energy harvesting, and dynamic environments testing; topics in modal analysis and parameter identification; and data science in engineering.

Each collection presents early findings from experimental and computational investigations on an important area within structural dynamics. Coupled structures or substructuring is one of these areas.

Substructuring is a general paradigm in engineering dynamics where a complicated system is analyzed by considering the dynamic interactions between subcomponents. In numerical simulations, substructuring allows one to reduce the complexity of parts of the system in order to construct a computationally efficient model of the assembled system. A subcomponent model can also be derived experimentally, allowing one to predict the dynamic behavior of an assembly by combining experimentally and/or analytically derived models. This can be advantageous for subcomponents that are expensive or difficult to model analytically. Substructuring can also be used to couple numerical simulation with real-time testing of components. Such approaches are known as hardware-in-the-loop or hybrid testing.

Whether experimental or numerical, all substructuring approaches have a common basis, namely the equilibrium of the substructures under the action of the applied and interface forces and the compatibility of displacements at the interfaces of the subcomponents. Experimental substructuring requires special care in the way the measurements are obtained and processed in order to assure that measurement inaccuracies and noise do not invalidate the results. In numerical approaches, the fundamental quest is the efficient computation of reduced-order models describing the substructure's dynamic motion. For hardware-in-the-loop applications, difficulties include the fast computation of the numerical components and the proper sensing and actuation of the hardware component. Current advances in experimental techniques, sensor/actuator technologies, novel numerical methods, and parallel computing have rekindled interest in substructuring in recent years leading to new insights and improved experimental and analytical techniques.

The organizers would like to thank the authors, presenters, session organizers, and session chairs for their participation in this track.

Provo, UT, USA — Matthew S. Allen

L'Aquila, Italy — Walter D'Ambrogio

Albuquerque, NM, USA — Dan Roettgen

Contents

Chapter 1
Sensitivity-Based Substructure Error Propagation for Efficient Assembly Model Reduction

B. M. Kessels, M. L. J. Verhees, A. M. Steenhoek, R. H. B. Fey, and N. van de Wouw

Abstract Due to ever-increasing complexity of structural dynamic systems in various fields of engineering, model reduction techniques using a substructuring approach, a.k.a. component mode synthesis techniques, still form an active field of research. This paper proposes an efficient, novel method for error approximation for model reduction of coupled substructures in structural dynamics. When coupling multiple reduced substructure models, the influence of individual substructure modes on the dynamic behavior of the total reduced mechanical assembly is generally unknown. Rather than selecting substructure eigenmodes, which are used to constitute the reduction bases, solely based on their eigenfrequencies, this paper proposes a different selection method. This method inspects the influence of individual substructure modes on an assembly receptance using so-called modal receptance error contributions. These modal receptance error contributions are defined as the assembly receptance reduction error induced by truncating individual substructure modes. By determining the sensitivity of the receptance of the assembly with respect to the uncoupled substructure receptances, a substructure reduction error is propagated through the assembly model, resulting in a first-order approximation of the assembly error. To calculate this sensitivity, the receptance of the assembly is expressed in terms of the individual receptances of the uncoupled substructures and Boolean mapping matrices, used to couple substructures. Comparing different modal receptance error contributions, associated with the reduction of individual substructures, provides insight in the selection of substructure modes which results in an efficient reduction of the assembly. As such, a mode-selection criterion is defined by using the obtained information on the sensitivity of the quality of the assembly reduction to truncating individual substructure modes. This criterion helps to determine a more efficient reduction basis. To illustrate the proposed method, a cantilever Euler beam consisting of two substructures is used.

Keywords Structural dynamics · Model reduction · CMS · Substructure mode selection · Error Propagation

1.1 Introduction

Modeling structural dynamics systems continues to be a challenging task due to the ever-increasing complexity of these systems. Furthermore, despite the advances in computer technology, this growing complexity makes computational work increasingly time consuming. One of the many fields where these problems are relevant is the semiconductor industry. For instance, lithographic systems, required for semiconductor production, demand extreme levels of accuracy and consequently detailed and accurate models. The finite element models required to achieve this accuracy generally have too many Degrees Of Freedom (DOFs) to be used within reasonable computational time. Therefore, the development of methodologies which reduce the size and complexity of these numerical models, while maintaining a desired level of accuracy, is still an active research field. A disadvantage of decreasing the number of DOFs is the resulting loss of the accuracy of the reduced model, which can be quantified by the difference between the responses and/or dynamic properties of the reduced model and similar

B. M. Kessels (✉) · R. H. B. Fey · N. van de Wouw
Eindhoven University of Technology, Department of Mechanical Engineering, Eindhoven, The Netherlands

University of Minnesota, Department of Civil, Environmental, and Geo-Engineering, Minneapolis, MN, USA
e-mail: b.m.kessels@tue.nl; r.h.b.fey@tue.nl; n.v.d.wouw@tue.nl

M. L. J. Verhees · A. M. Steenhoek
ASML, System Dynamics, Veldhoven, The Netherlands
e-mail: thijs.verhees@asml.com; alexander.steenhoek@asml.com

M. S. Allen et al. (eds.), *Dynamic Substructures, Volume 4*, Conference Proceedings of the Society for Experimental Mechanics Series, https://doi.org/10.1007/978-3-030-75910-0_1

quantities of the unreduced model. Therefore, reduction methods have been devised which try to optimize the trade-off between improvement of the efficiency and loss of accuracy.

An example of such reduction methods is moment matching [1]. Since this method is applied in the frequency domain, useful insights as would be obtained in the time domain (e.g., eigenvalues with corresponding mode shapes of reduced structures) are lost though. Two other well-established reduction methods are balanced truncation [1] and Hankel norm approximation [8], which both focus on maintaining accurate input–output behavior. Although both methods provide useful a priori bounds on the reduction error, the first-order description of the system used for these methods loses physical interpretability after reduction. A fourth class of reduction methods which does not suffer from loss of physical insight are Component Mode Synthesis (CMS) methods [2], in which components are reduced by truncating component vibration modes, targeting on accuracy in a certain range of frequencies.

To manage the complexity of models, they are often divided into multiple substructures or components which can be modeled and analyzed separately. This approach is referred to as Dynamic Substructuring (DS) [10]. Additionally, the number of DOFs of each substructure model can be decreased to reduce computational costs. Subsequently, the reduced substructure models are coupled to obtain a reduced assembly model. The methods belonging to the aforementioned class of CMS methods use the concept of DS. Often, static (correction) modes are involved in the process of coupling substructure models, i.e., realizing compatibility of neighboring interface DOFs and achieving internal interface force equilibrium. Therefore, these static modes are typically present in the reduction bases of CMS methods. Additionally, typically, a selection of other mode-types, e.g., retained (or non-truncated) eigenmodes, is present in these reduction bases as well. Depending on the specific CMS method, the types of employed eigenmodes vary. For example, the Craig–Bampton method [3] uses eigenmodes of a substructure which is virtually constrained at its interfaces with other substructures. Oppositely, the Hintz–Herting (HH) method [9], for instance, uses eigenmodes which are free at the interfaces.

The truncation of eigenmodes is usually done by selecting eigenmodes based on their eigenfrequencies. Since high-frequency behavior is typically of less interest than the behavior at low frequencies, eigenmodes with low eigenfrequencies are usually retained, whereas high-frequency eigenmodes are truncated [7, 9, 10]. Although this approach is often viable, the set of "kept" eigenmodes might not be optimal when multiple substructures are coupled. This statement is motivated by the fact that some (local) component eigenmodes, or even entire substructures, may have very little influence on the behavior of (relevant) assembly transfer functions. Hence, low-frequency component eigenmodes, which have only little influence on the relevant assembly behavior, are unnecessarily retained, leaving room for improvement. In addition, high-frequency component eigenmodes that show relevant (local) behavior might be truncated. Techniques that evaluate the influence of component modes on the assembly transfer function in order to select which modes should be retained are however scarce. More specifically, techniques that select component modes such that specific error measures of reduced assembly transfer functions are minimized seem to be missing in literature. Therefore, this paper proposes a novel method which, based on a sensitivity analysis, selects substructure eigenmodes resulting in an accurate and efficient reduction basis for each component. As will be shown, this method yields reduced assembly models with lower reduction errors compared to assembly models which are reduced using the traditional component eigenmode-selection method, i.e., using eigenfrequencies. Since the method presented here requires free-interface eigenmodes, the previously mentioned Hintz–Herting CMS method is adopted for this research.

The proposed method is based on the propagation of substructure errors which is efficiently approximated by exploiting a sensitivity-based description of the assembly reduction receptance error as explained in Sect. 1.2. This section also discusses an approximation of the substructure receptance error which improves the efficiency of the selection procedure. Subsequently, Sect. 1.3 explains how the estimated assembly receptance error is used to select substructure eigenmodes. Results obtained for a case study, where a beam system consisting of two substructures is used, are then presented in Sect. 1.4. Finally, the conclusions and some suggestions for future work are listed in Sect. 1.5.

1.2 Sensitivity-Based Substructure Error Propagation

In this section, an efficient approach to estimate the error in the assembly receptance originating from substructure reduction errors is introduced. To achieve this, the assembly receptance error is described in terms of individual contributions of substructure receptance errors using a first-order approximation [5]:

$$\boldsymbol{\Delta Y}^{(assy)}(j\omega) = \sum_{k}^{n_{uc}} \sum_{l}^{n_{uc}} \frac{\partial \boldsymbol{Y}^{(assy)}}{\partial \boldsymbol{Y}[k,l]}(\omega j)\boldsymbol{\Delta Y}^{(uc)}[k,l](j\omega). \tag{1.1}$$

Here, matrix $\boldsymbol{\Delta Y}^{(assy)}$ denotes the approximated error in the assembly receptance matrix $\boldsymbol{Y}^{(assy)}$, n_{uc} denotes the combined number of all DOFs of all uncoupled components, and ω denotes the angular eigenfrequency. This error consists of the product of two multiplication factors: (i) the sensitivity of the assembly with respect to a change in a single component ($\frac{\partial \boldsymbol{Y}^{(assy)}}{\partial \boldsymbol{Y}[k,l]}$), and (ii) the receptance error induced in the uncoupled components ($\boldsymbol{\Delta Y}^{(uc)}[k,l]$), where $\boldsymbol{\Delta Y}^{(uc)}$ contains the uncoupled component receptance error matrices $\boldsymbol{\Delta Y}^{(s)}$ on its diagonal, and $[k,l]$ selects the matrix-element in row k and column l. These terms are derived separately in the following exposition, starting with the former.

1.2.1 Sensitivity of Assembly with Respect to a Substructure

The first multiplication factor in the right-hand side of (1.1) serves to separate the contributions of individual substructures to the assembly error. To isolate these contributions, the Lagrange Multiplier Frequency-Based Substructuring (LM FBS) framework is adopted [6]. This framework defines the assembly receptance matrix as

$$\boldsymbol{Y}^{(assy)} = \boldsymbol{Y} - \boldsymbol{Y}\boldsymbol{Q}\boldsymbol{Y}, \tag{1.2}$$

where

$$\boldsymbol{Q} = \boldsymbol{B}_f^T \left(\boldsymbol{B}_f \boldsymbol{Y} \boldsymbol{B}_f^T\right)^{-1} \boldsymbol{B}_f. \tag{1.3}$$

Here, $\boldsymbol{Y}$ is a block diagonal matrix containing the uncoupled substructure receptance matrices in the blocks on its diagonal and $\boldsymbol{B}_f$ is a signed Boolean mapping matrix ensuring compatibility at the substructure interfaces. Exploiting the linearity of a matrix with respect to its elements, the partial derivative of (1.2) with respect to $\boldsymbol{Y}[k,l]$ is efficiently calculated:

$$\frac{\partial \boldsymbol{Y}^{(assy)}}{\partial \boldsymbol{Y}[k,l]} = \boldsymbol{P}_{kl} - \boldsymbol{P}_{kl}\boldsymbol{Q}\boldsymbol{Y} + \boldsymbol{Y}\boldsymbol{Q}\boldsymbol{P}_{kl}\boldsymbol{Q}\boldsymbol{Y} - \boldsymbol{Y}\boldsymbol{Q}\boldsymbol{P}_{kl}, \tag{1.4}$$

where $\boldsymbol{P}_{kl}$ is a single-entry matrix with $\boldsymbol{P}_{kl}[k,l] = 1$. For a detailed derivation of (1.4), the reader is referred to [4].

1.2.2 Error in Uncoupled Substructures

The second multiplication factor in the right-hand side of (1.1) is the (estimated) reduction receptance error within a substructure. Since these reduced substructures are to be coupled, CMS is used for the reduction [10]. Although (1.1) contains multiple substructure receptance error matrices contained in a block diagonal matrix ($\boldsymbol{\Delta Y}^{(uc)}$), this section focuses on the computation of a single reduction receptance error matrix of substructure s ($\boldsymbol{\Delta Y}^{(s)}$), where the explicit dependency on s is omitted for clarity in the following. One method to achieve this reduction is the aforementioned Hintz–Herting CMS method [9], in which the reduction matrix is defined as

$$\boldsymbol{T}_{HH} = \left[\boldsymbol{\Phi}_{\mathcal{K}}\ \boldsymbol{\Phi}_{\text{ir}}\ \boldsymbol{\Psi}_{\text{const}}\right] =: \left[\boldsymbol{\Phi}_{\mathcal{K}}\ \boldsymbol{X}\right]. \tag{1.5}$$

Here, $\boldsymbol{\Phi}_{\text{ir}}$ represents a set containing n_{ir} inertia relief modes and $\boldsymbol{\Psi}_{\text{const}}$ is a set consisting of n_b constraint modes which ensure statically exact behavior at the boundary nodes [9]. The set of boundary nodes is constituted by the nodes at the interfaces and nodes at which external forces/moments are applied. Furthermore, $\boldsymbol{\Phi}_{\mathcal{K}}$ represents a set of $n_{\mathcal{K}} \leq n_{DOF} - n_{ir} - n_b$ kept free-interface eigenmodes, where n_{DOF} represents the number of DOFs in the substructure. Although typically the eigenmodes with the lowest associated eigenfrequencies are used to constitute this set, other combinations of (kept) eigenmodes can be used as well. Note that in (1.5), in contrast to the definition found in [9], $\boldsymbol{\Phi}_{\text{ir}}$ and $\boldsymbol{\Psi}_{\text{const}}$ are orthogonalized with respect to $\boldsymbol{\Phi}_{\mathcal{K}}$. This orthogonality ensures that the reduced mass, stiffness, and, when modal damping is assumed, damping matrices are of a block diagonal nature. Consequently, a (partial) spectral decomposition of the HH reduced receptance matrix of a substructure is acquired:

$$\hat{\boldsymbol{Y}}_{HH}(j\omega) = \boldsymbol{Y}_{\boldsymbol{X}}(j\omega) + \sum_{k\in\mathcal{K}} \boldsymbol{Y}_{\boldsymbol{\phi}_k}(j\omega) \\ =: \boldsymbol{Y}_{\boldsymbol{X}}(j\omega) + \hat{\boldsymbol{Y}}_{\mathcal{K}}(j\omega), \tag{1.6}$$

where $\mathcal{K}$ represents the set of mode numbers related to the set of kept modes. Furthermore, the receptance contribution of the inertia relief and constraint modes is defined as

$$\boldsymbol{Y}_{X}(j\omega) = \boldsymbol{X}\left(\hat{\boldsymbol{K}}_X + j\omega\hat{\boldsymbol{D}}_X - \omega^2\hat{\boldsymbol{M}}_X\right)^{-1}\boldsymbol{X}^T, \tag{1.7}$$

which, if $n_{ir} + n_b$ is small, uses a relatively inexpensive inverse, and where

$$\hat{\boldsymbol{M}}_X = \boldsymbol{X}^T\boldsymbol{M}\boldsymbol{X}, \qquad \hat{\boldsymbol{D}}_X = \boldsymbol{X}^T\boldsymbol{D}\boldsymbol{X}, \qquad \hat{\boldsymbol{K}}_X = \boldsymbol{X}^T\boldsymbol{K}\boldsymbol{X}. \tag{1.8}$$

Alternatively, superposition of modal contributions to $\boldsymbol{Y}_X$ based on an eigenvalue problem constituted by $\hat{\boldsymbol{M}}_X$, $\hat{\boldsymbol{D}}_X$, and $\hat{\boldsymbol{K}}_X$ could be employed to calculate $\boldsymbol{Y}_X$. Furthermore, in case of mass-normalized eigenmodes, the modal substructure receptance contributions in (1.6) are given by

$$\boldsymbol{Y}_{\boldsymbol{\phi}_k}(j\omega) = \frac{\boldsymbol{\phi}_k\boldsymbol{\phi}_k{}^T}{\omega_k{}^2 - 2j\xi_k\omega_k\omega - \omega^2}. \tag{1.9}$$

Note that, if $n_{\mathcal{K}} = n_{\mathrm{DOF}} - n_{\mathrm{ir}} - n_{\mathrm{b}}$, the system is not reduced and thus the receptance matrix $\hat{\boldsymbol{Y}}_{HH}$ does actually not represent a reduced receptance and is therefore equivalent to the unreduced receptance matrix. Since the aim of the method proposed in this paper is to select which eigenmodes in the set $\mathcal{K}$ should be retained such that an accurate reduced assembly receptance is obtained, the influence of each eigenmode should be investigated. This is achieved by evaluating, for each eigenmode, what the receptance error would be if exclusively that eigenmode is truncated from the set $\mathcal{K}$, such that a set $\mathcal{K}_{\mathrm{eval}}$ containing $n_{\mathcal{K}} - 1$ eigenmodes is obtained. In a practical setting, it is however often not feasible to compute all eigenmodes of a large system. The set $\mathcal{K}$ may therefore denote a limited set of eigenmodes obtained by a coarse modal truncation, in which a relatively large set of eigenmodes with the lowest eigenfrequencies are retained. In this case, the proposed eigenmode-selection method will select a subset of this large set $\mathcal{K}$ to form the new smaller set $\mathcal{K}_{\mathrm{sel}}$ containing $n_{\mathrm{sel}} < n_{\mathcal{K}}$ eigenmodes.

To avoid calculating $\boldsymbol{Y}_X(j\omega)$, which depends on the specific set of kept modes, for each evaluated set $\mathcal{K}_{\mathrm{eval}}$ such that the corresponding HH receptance reduction error is obtained, the Modal Truncation (MT) receptance error is used instead. This error is computed relatively cheaply for multiple small sets of individually truncated eigenmodes using modal superposition:

$$\boldsymbol{\Delta Y} = \sum_{k\in\mathcal{D}_{\mathrm{eval}}} \boldsymbol{Y}_{\boldsymbol{\phi}_k}(j\omega) = \sum_{k\in\mathcal{D}_{\mathrm{eval}}} \frac{\boldsymbol{\phi}_k\boldsymbol{\phi}_k{}^T}{\omega_k{}^2 - 2j\xi_k\omega_k\omega - \omega^2}. \tag{1.10}$$

Here, $\boldsymbol{\phi}_k$, ω_k, and ξ_k denote the kth eigenmode, kth angular eigenfrequency, and kth modal damping factor, respectively. Furthermore, $\mathcal{D}_{\mathrm{eval}}$ represents the set of deleted (or truncated) eigenmodes, which in the proposed methodology thus contains only a single eigenmode for each evaluation. Note that, as shown in (1.6), the HH reduced receptance ($\hat{\boldsymbol{Y}}_{HH}$) is not exactly equal to the MT reduced receptance ($\hat{\boldsymbol{Y}}_{\mathcal{K}}$) due to the influence of $\boldsymbol{Y}_X(j\omega)$. Nevertheless, it is expected that an "optimal" selection of eigenmodes $\mathcal{K}_{\mathrm{sel}}$, obtained using an analysis based on the MT receptance error, is a valid approximation of the "optimal" set of eigenmodes for a HH reduction basis. This is motivated by the fact that the addition of $\boldsymbol{Y}_X(j\omega)$ in (1.6) always improves the accuracy of $\hat{\boldsymbol{Y}}_{\mathcal{K}}$. Therefore, in order to achieve an "optimal" $\hat{\boldsymbol{Y}}_{HH}$, it is assumed that it is best practice to first find the "optimal" $\hat{\boldsymbol{Y}}_{\mathcal{K}}$. Accordingly, (1.10) is substituted, together with (1.4), in (1.1) to estimate the assembly receptance reduction error caused by truncating a single substructure eigenmode. A methodology which uses this approximation to determine which sets of substructure eigenmodes constitute substructure MT reduction bases that minimize the assembly receptance error is introduced in the following section. The sets of eigenmodes found using this methodology are then used in combination with the constraint and inertia relief modes in HH reduction bases to reduce the substructures which are eventually coupled to obtain the reduced assembly model.

1.3 Component Mode Selection

In this section, the approximated assembly receptance reduction error is used to directly assess the influence of individual substructure eigenmodes on the assembly receptance reduction error. For this assessment, a leave-one-out approach is adopted where the (additional) assembly receptance error, caused by truncation of a single eigenmode, is estimated for each substructure eigenmode and in all substructures. Although the error of each assembly Frequency Response Function (FRF) related to all DOFs in the receptance matrix can be taken into account, often only a limited number of assembly FRFs, related to the n_{POIs} Points Of Interest (POIs), is relevant. Here, the latter approach will be adopted. Consequently, the estimated receptance reduction error of these relevant FRFs are stored in a (square) matrix $\boldsymbol{\Delta Y}_{\text{POIs}}^{(assy)}$. This receptance error matrix is a function of the frequency. Therefore, first, spatial norms of the assembly receptance error matrix are calculated at each evaluated frequency point, resulting in a vector where each entry corresponds to a single frequency point. Subsequently, a "temporal" norm is used such that this vector is condensed to a single scalar error quantity. These scalar values (obtained by truncating different substructure eigenmodes) make comparing errors significantly more straightforward and suitable for automation. The procedure of taking spatial and temporal norms is schematically summarized in Fig. 1.1.

In this work, the Frobenius norm is used as spatial norm since it takes the value of each matrix-entry into account with equal weight (note that a separate weighting matrix can however be added if required):

$$\hat{\epsilon}^{(assy)}(\omega) = \frac{\left\| \boldsymbol{\Delta Y}_{\text{POIs}}^{(assy)}(j\omega) \right\|_F}{\left\| \boldsymbol{Y}_{\text{POIs}}^{(assy)}(j\omega) \right\|_F}. \tag{1.11}$$

Here, the hat on top of ϵ represents the approximated nature of this (frequency-dependent) error measure. Additionally, as shown in (1.11), the Frobenius norm of the receptance error is divided by the Frobenius norm of the unreduced receptance matrix such that a relative norm is obtained. This choice is motivated by the assumption that the response at all frequencies should be approximated equally well in a relative sense. A temporal norm, specifically the 2-norm, is then used to condense the frequency-dependent error measure $\hat{\epsilon}^{(assy)}(\omega)$ into a single scalar error quantity:

$$\hat{\chi}^{(assy)} = \left\| W(\omega)\hat{\epsilon}^{(assy)}(\omega) \right\|_2, \tag{1.12}$$

where $W(\omega)$ represents a frequency-filtering function. In this paper, brick-wall filtering is used to exclusively evaluate errors within a specified Frequency Range Of Interest (FROI).

As mentioned before, a leave-one-out approach is adopted to establish the set of kept substructure eigenmodes. For each free-interface eigenmode k of each substructure s, the error measure $\hat{\chi}_{\boldsymbol{\phi}_k^{(s)}}^{(assy)}$ is calculated as if mode k of component s would be the only mode to be truncated in the entire assembly. Afterwards, all error measures of all substructures are sorted by magnitude. The eigenmodes related to the $n_{\mathcal{K}}^{(assy)}$ highest error measures are then selected as (kept) vibration modes in the new (HH) reduction bases of each substructure. Here, $n_{\mathcal{K}}^{(assy)}$ can be chosen a priori by the user based on required model sizes (in terms of, e.g., number of DOFs). Alternatively, the choice of $n_{\mathcal{K}}^{(assy)}$ can be based on a required (maximal) error level.

Finally, note that the set of eigenmodes, found using (estimated) receptance errors, will be equivalent to the set of eigenmodes that is found based on (estimated) mobility and inertance errors. This statement is using the fact that a time derivative in the frequency domain is a multiplication with $j\omega$ which is negated by the relative norm in (1.11).

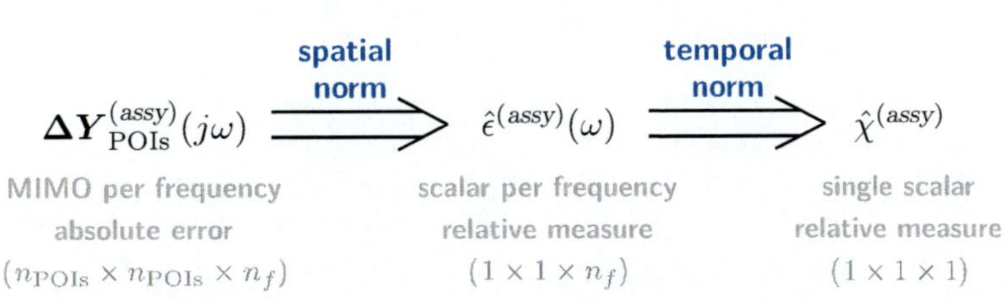

Fig. 1.1 Schematic overview of the procedure to determine the relative error measure by using spatial and temporal norms. Indicated between brackets are the dimensions of each variable/measure (n_{POIs} is the number of POIs and n_f is the number of evaluated frequency points)

1.4 Results

1.4.1 Demonstrator Description

To illustrate the method introduced in the previous sections, a simple Euler–Bernoulli cantilever beam is used. As shown in Fig. 1.2, this beam is cut in half to obtain two substructures. The material and geometry parameters of both substructures are listed in Table 1.1 (note the relatively low Young's modulus for substructure A). Additionally, proportional modal damping is added to each free-interface eigenmode of both substructures with $\xi = 0.05$ for all eigenmodes. To model the substructures, they are divided into 10 identical linear 2-node Euler beam elements each, resulting in 11 nodes per substructure, where each node has one transversial displacement and one rotational DOF. Consequently, the unreduced models of components A and B have 20 and 22 DOFs, respectively, and the unreduced assembly model has 40 DOFs. As indicated in Fig. 1.2, one node per substructure is required for coupling and is labeled as interface node. These interface nodes are part of the set of boundary nodes. Other nodes belonging to the set of boundary nodes are the nodes at which external forces/moments are applied. These nodes are also referred to as retained nodes. Since these boundary nodes represent relevant nodes in the system, they are also referred to as points of interest. In this example, only the transfer functions between the forces and moments, and displacements and rotations at the POIs are considered relevant. Therefore, $\boldsymbol{Y}^{(assy)}_{\text{POIs}}$ and $\boldsymbol{\Delta Y}^{(assy)}_{\text{POIs}}$ exclusively contain the FRFs related to DOFs of the boundary nodes.

1.4.2 Ranking of Substructure Eigenmodes

As discussed earlier, the first step in the mode-selection procedure is to estimate the assembly errors caused by truncating individual modes, $\boldsymbol{\Delta Y}^{(assy)}_{\boldsymbol{\phi}_k^{(s)}}$, using (1.1). Then, the estimated error measures $\hat{\chi}^{(assy)}_{\boldsymbol{\phi}_k^{(s)}}$ are obtained by first applying (1.11) to $\boldsymbol{\Delta Y}^{(assy)}_{\boldsymbol{\phi}_k^{(s)}}$ and then using (1.12). Both the frequency-dependent error measures based on the spatial norm, i.e., $\hat{\epsilon}^{(assy)}(\omega)$, and the scalar error measures, i.e., $\hat{\chi}^{(assy)}$, are plotted in Fig. 1.3 for the first two free-interface eigenmodes of both substructures. As illustrated by this figure, the error measures for the eigenmodes of substructure A are clearly higher than those of substructure B. Note that there is a large difference in $\hat{\epsilon}^{(assy)}$ at low frequencies. Therefore, these eigenmodes of substructure A should be kept in the reduced model of substructure A, since they are expected to cause relatively large errors, if truncated, compared to the examined eigenmodes of substructure B.

In Table 1.2, the scalar error measures related to all (elastic free-interface) eigenmodes of both substructures, obtained for a FROI $= 0 - 32$ Hz, are listed. This table also specifies the eigenfrequency of each eigenmode. Note that, due to the use of the MT receptance error approximation as introduced in Section II, there is no limitation on the number of evaluated

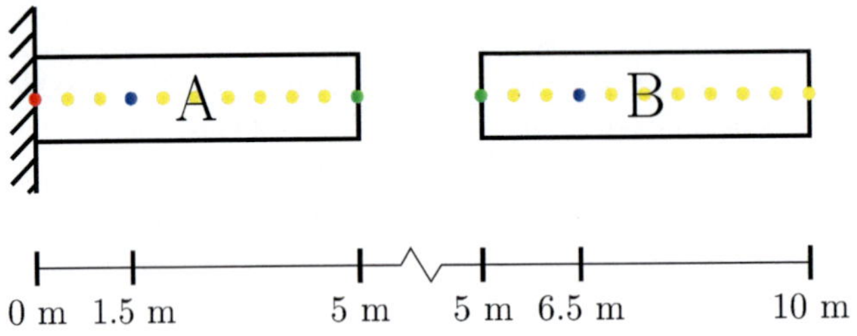

Fig. 1.2 Schematic representation of a two-substructure cantilever beam. The fixed, interface, retained, and internal nodes of each substructure are indicated by the red, green, blue, and yellow dots, respectively

Table 1.1 Material and geometry parameters for the two-substructure cantilever beam

Parameter	Unit	Substructure A	Substructure B
Density	kg/m^3	7.850×10^3	7.850×10^3
Cross-sectional area	m^2	1×10^{-2}	1×10^{-2}
Area moment of inertia	m^4	$1/12 \times 10^{-4}$	$1/12 \times 10^{-4}$
Young's modulus	N/m^2	2.1×10^9	2.1×10^{11}

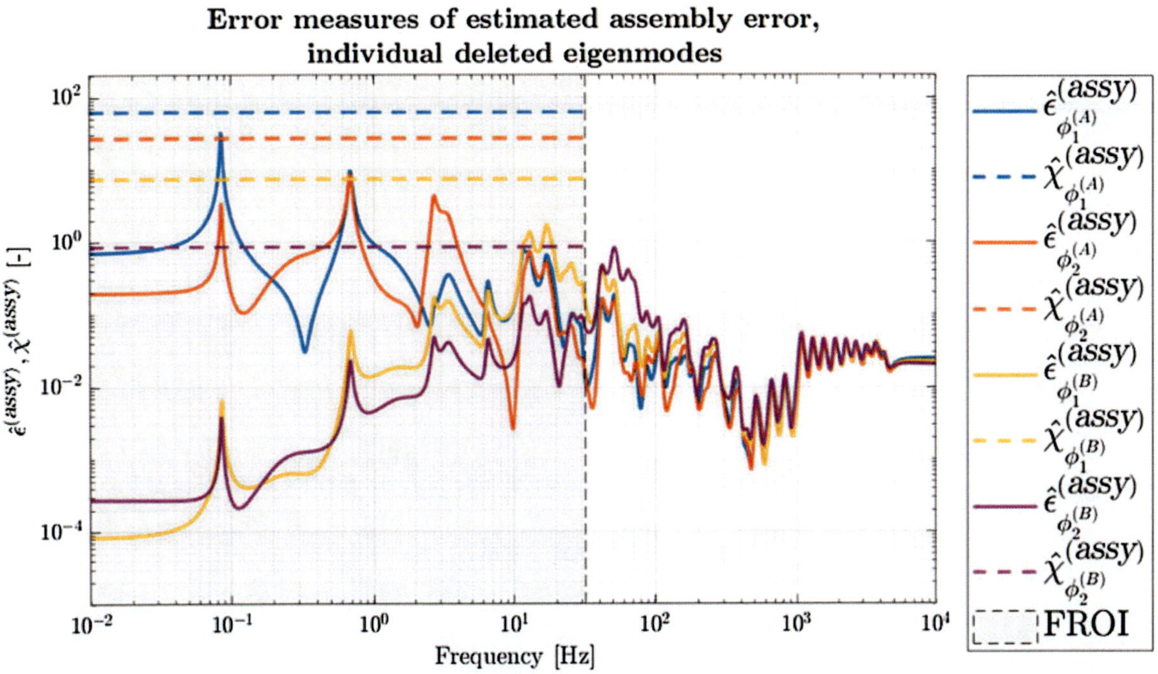

Fig. 1.3 Approximated assembly receptance error measures $\hat{\epsilon}^{(assy)}(\omega)$ and $\hat{\chi}^{(assy)}$, obtained by deleting specific substructure free-interface eigenmodes, with FROI $= 0 - 32$ Hz

Table 1.2 Eigenfrequencies and estimated scalar error measures for the assembly receptance error when only a single substructure elastic free-interface eigenmode is truncated, listed for every eigenmode of both substructures of the demonstrator beam. The kth eigenfrequency of unreduced substructure s is denoted by $f_k^{(s)}$

	Substructure A		Substructure B			Substructure A		Substructure B	
Mode k	$f_k^{(A)}$ [Hz]	$\hat{\chi}^{(assy)}_{\phi_k^{(A)}}$ [−]	$f_k^{(B)}$ [Hz]	$\hat{\chi}^{(assy)}_{\phi_k^{(B)}}$ [−]	Mode k	$f_k^{(A)}$ [Hz]	$\hat{\chi}^{(assy)}_{\phi_k^{(A)}}$ [−]	$f_k^{(B)}$ [Hz]	$\hat{\chi}^{(assy)}_{\phi_k^{(B)}}$ [−]
1	0.334	61.790	21.267	7.358	11	114.735	2.389	1372.920	0.036
2	2.095	26.832	58.636	0.854	12	138.614	1.413	1657.922	0.023
3	5.866	9.149	115.028	0.405	13	168.032	0.681	1992.830	0.028
4	11.503	8.964	190.428	0.238	14	202.804	1.353	2384.978	0.028
5	19.045	12.233	285.225	0.115	15	243.699	1.347	2843.043	0.018
6	28.532	3.188	400.022	0.081	16	291.466	1.022	3370.610	0.020
7	40.031	2.407	535.536	0.072	17	345.925	0.559	3949.969	0.024
8	53.631	4.835	691.699	0.046	18	403.850	1.030	4499.502	0.012
9	69.344	1.326	859.280	0.030	19	454.780	1.156	5686.193	0.058
10	86.205	0.713	1141.111	0.038	20	569.152	3.381	5696.749	0.057

eigenmodes due to concerns related to orthogonality of the HH reduction basis. Since for the demonstrator used here, all n_{DOF} eigenmodes can be calculated, each eigenmode of both substructures is evaluated. The error measures in Table 1.2 are sorted by magnitude, resulting in Table 1.3, where the 13 eigenmodes with the highest error measures are ranked (high to low) in the middle row. Additionally, the eigenmodes are ranked by eigenfrequency (low to high) in the last row as is done in the traditional eigenmode-selection procedure. As is shown in Table 1.3, some differences can be observed between the ranking of eigenmodes using the traditional method and using the newly proposed, or sensitivity-based, method. For example, according to the sensitivity-based method, the fifth eigenmode of substructure A is the third kept mode of choice, whereas it is the fifth mode of choice based on the eigenfrequencies. Also, the 8th eigenmode of substructure A is expected to be more relevant than modes 6 and 7 of substructure A according to the error measure-based ranking compared to the eigenfrequency-based ranking.

Now, using Table 1.3, substructure eigenmodes can be selected such that an accurate and efficient reduced assembly model is obtained. For the traditional method (using the cutoff frequencies of the component eigenmodes) the first $n_{\mathcal{K}}^{(assy)}$ eigenmodes in the "traditional method" row are selected, since these correspond to the eigenmodes with the lowest

Table 1.3 Ranking of the 13 most relevant elastic free-interface eigenmodes of both substructures, based on the newly proposed (sensitivity-based) and traditional (cutoff frequency) eigenmode-selection method. The error measures are determined for a FROI $= 0 - 32$ Hz

Ranking of substructure eigenmodes		1	2	3	4	5	6	7	8	9	10	11	12	13	...
Proposed method	Substructure	A	A	A	A	A	B	A	A	A	A	A	A	A	
	Mode	1	2	5	3	4	1	8	20	6	7	11	12	14	
Traditional method	Substructure	A	A	A	A	A	B	A	A	A	B	A	A	A	
	Mode	1	2	3	4	5	1	6	7	8	2	9	10	11	

eigenfrequencies. Oppositely, for the sensitivity-based method, the first $n_{\mathcal{K}}^{(assy)}$ eigenmodes in the "proposed method" row are selected, because these are expected to generate the highest assembly reduction error (measure) when truncated. Note that the different order of eigenmodes for both methods suggests that the traditional method is not optimal for the defined error measure.

In part C of this section, see below, three examples will be given in which the newly proposed selection method outperforms the traditional method. To compare both methods, the exact assembly receptance error will be used:

$$\boldsymbol{E}_{\text{POIs}}^{(assy)}(j\omega) = \boldsymbol{Y}_{\text{POIs}}^{(assy)}(j\omega) - \hat{\boldsymbol{Y}}_{\text{POIs}}^{(assy)}(j\omega). \tag{1.13}$$

In (1.13), $\hat{\boldsymbol{Y}}_{\text{POIs}}^{(assy)}$ is the reduced assembly receptance matrix obtained by reducing the substructures with a set of kept eigenmodes (as selected using Table 1.3), where only the FRFs corresponding to the POIs are contained within this matrix. To make this comparison fair and straightforward, similar error measures as used for the estimated receptance error measure calculation ((1.11) and (1.12)) are determined for the reduced assembly. Since now the assembly error is not estimated but exactly calculated for the sake of the comparison, the exact receptance error measures are denoted without hat, i.e., as $\epsilon^{(assy)}(\omega)$, and $\chi^{(assy)}$. To this end, the estimated assembly error, $\boldsymbol{\Delta Y}_{\text{POIs}}^{(assy)}$, in (1.11) is replaced by the exact assembly receptance error, $\boldsymbol{E}_{\text{POIs}}^{(assy)}$.

1.4.3 Comparison of Different Sets of Kept Eigenmodes

In this section, the exact assembly reduction errors as obtained by using the traditional and improved eigenmode-selection procedures are compared for 3 use cases.

In the first use case, it is assumed that the user only wants to retain 3 elastic free-interface eigenmodes for the Hintz–Herting reduction basis. Note that besides these eigenmodes, the required constraint and inertia relief modes are also present in the substructure reduction bases. Note that, in this example, the constraint modes of substructure B represent the rigid-body modes. Using the traditional selection method, solely elastic free-interface eigenmodes 1, 2, and 3 of substructure A would be kept in the reduction basis of substructure A, as indicated in Table 1.3. In contrast, the sensitivity-based method selects eigenmode 5 instead of 3 of substructure A. In Fig. 1.4, the associated exact receptance error measures of the resulting reduced assembly models are compared. As is shown in Fig. 1.4, $\epsilon^{(assy)}(\omega)$ indicates that the proposed method results in a favorable frequency-dependent error measure, especially near the end of the FROI. After taking the temporal norm, $\chi^{(assy)}$ is shown to be the lowest for the sensitivity-based method, as is also demonstrated in Table 1.5. This indicates that, in this case, with the same number of DOFs in the reduced model, the sensitivity-based selection method results in a more accurate reduced assembly receptance.

Since the first use case selects a relatively low amount of eigenmodes (note that the highest eigenfrequency of the 3 kept elastic free-interface eigenmodes is below the highest frequency in the FROI), for the second use case a larger set of eigenmodes is selected. More specifically, the number of eigenmodes is based on a rule of thumb often used in practice where the cutoff frequency used to reduce the substructures (by applying CMS) is approximately equal to 2.5 times the upper bound of the FROI. For the FROI investigated here, this implies that all eigenmodes with an eigenfrequency up to 80 Hz should be retained. Since, according to Table 1.2, there are in total 11 eigenmodes in both substructures with an eigenfrequency below 80 Hz, these eigenmodes are selected for the traditional approach. For a fair comparison in terms of computation time, the same number of eigenmodes is selected using Table 1.3 for the sensitivity-based method. As shown in Fig. 1.5, the sensitivity-based method results again in a lower receptance error measure, indicating that, also for this use case, the proposed

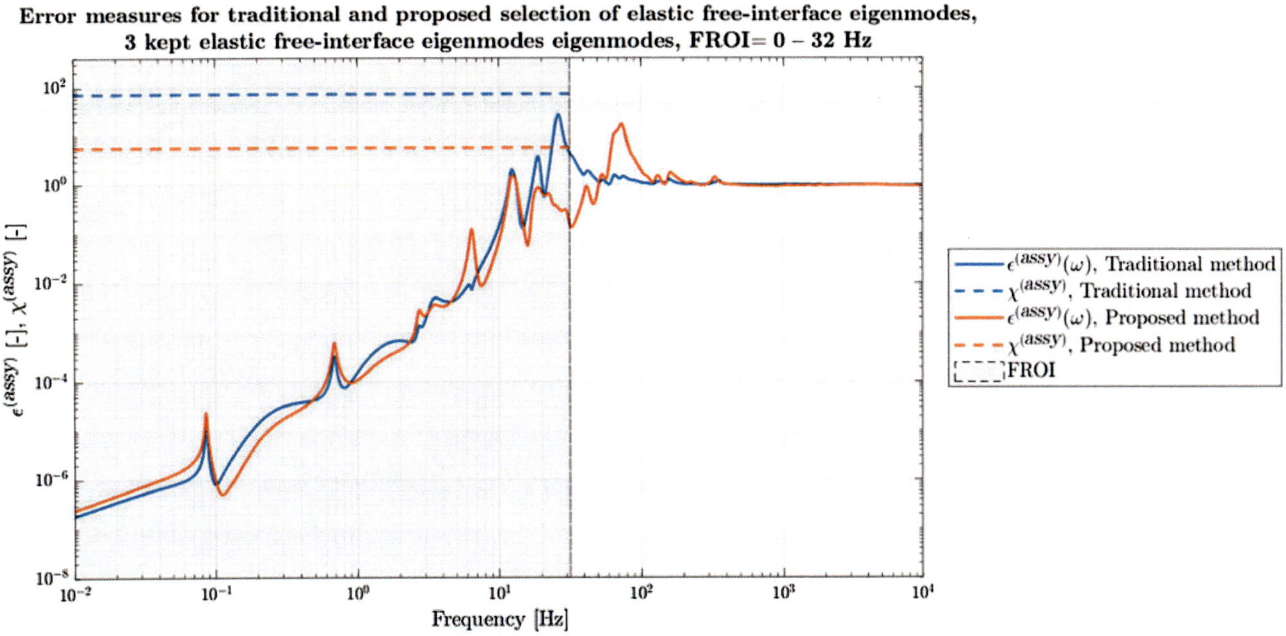

Fig. 1.4 Assembly receptance error measures $\epsilon^{(assy)}(\omega)$ and $\chi^{(assy)}$ obtained by using HH reduction bases where 3 elastic free-interface substructure eigenmodes are selected using the traditional method and the proposed method

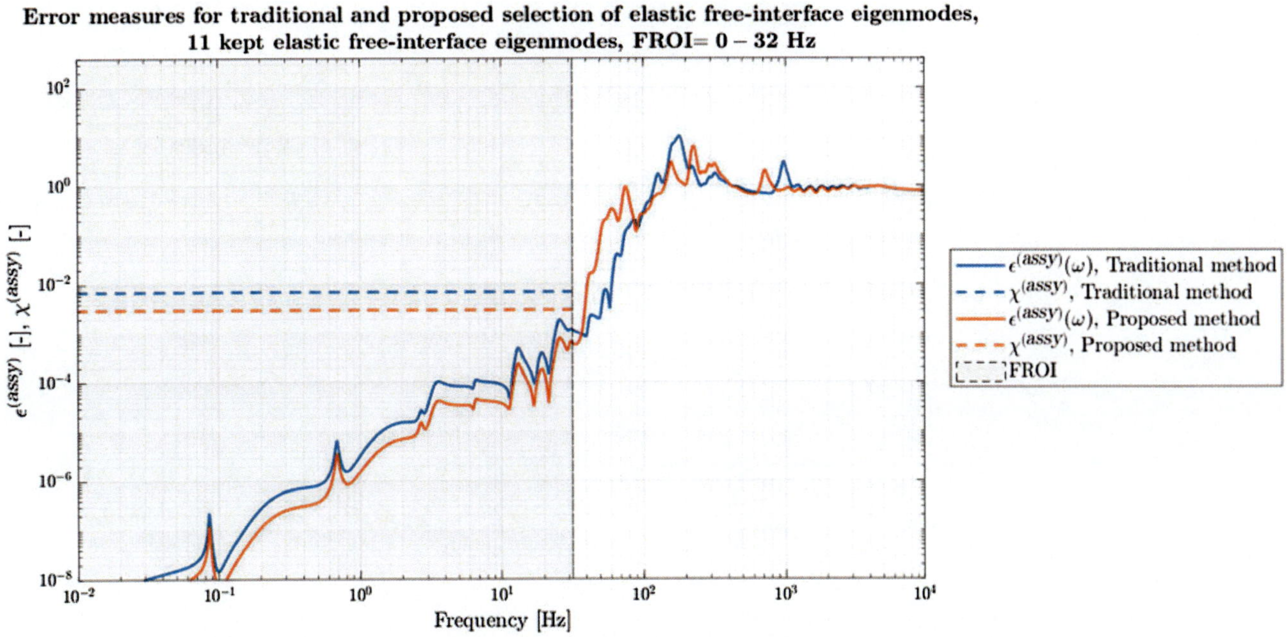

Fig. 1.5 Assembly receptance error measures $\epsilon^{(assy)}(\omega)$ and $\chi^{(assy)}$ obtained by using HH reduction bases where 11 elastic free-interface substructure eigenmodes are selected using the traditional method and the proposed method

method outperforms the traditional method. Furthermore, Fig. 1.5 also shows that the frequency-dependent receptance error measure of the sensitivity-based method is below that of the traditional method in the entire FROI.

Additionally, the relative errors of the eigenfrequencies of the reduced assembly models obtained using both selection methods with respect to the unreduced assembly eigenfrequencies, are shown in Table 1.4. Here it is shown that, although all errors are negligibly small, the relative eigenfrequency errors obtained using the traditional method are smaller than those obtained using the proposed method for almost every eigenfrequency in the FROI. The discrepancy in relative eigenfrequency errors between both methods grows in the favor of the traditional method for increasing eigenfrequencies, which, although

Table 1.4 Eigenfrequencies of the unreduced assembly model and relative eigenfrequency errors of assembly models reduced using the proposed and traditional eigenmode-selection methods to retain 11 elastic free-interface substructure eigenmodes. Only the eigenfrequencies within the FROI are shown

Assembly eigenmode k	1	2	3	4	5	6	7	8
Eigenfrequency, unreduced [Hz]	0.0857058	0.685552	2.70482	6.39387	11.6312	17.0967	21.8219	29.9792
Relative error, proposed method [%]	1.27×10^{-6}	8.73×10^{-10}	-9.91×10^{-8}	-3.54×10^{-6}	-4.71×10^{-5}	-1.88×10^{-4}	-2.94×10^{-4}	-4.03×10^{-3}
Relative error, traditional method [%]	1.29×10^{-6}	1.31×10^{-9}	-3.25×10^{-8}	-8.26×10^{-7}	-5.05×10^{-6}	-4.40×10^{-6}	-1.06×10^{-4}	-3.39×10^{-4}

Table 1.5 Exact scalar assembly error measures obtained when the reduction is performed by selecting $n_{\mathcal{K}}^{(assy)}$ elastic free-interface substructure eigenmodes for HH reduction bases, using the traditional (cutoff frequency) and proposed (sensitivity-based) eigenmode-selection methods

$n_{\mathcal{K}}^{(assy)}$	$\chi^{(assy)}$ [−] Traditional	Sensitivity-based
3	7.3343×10^{1}	5.7036×10^{0}
11	7.0686×10^{-3}	3.0286×10^{-3}
13	3.0401×10^{-3}	1.5924×10^{-3}

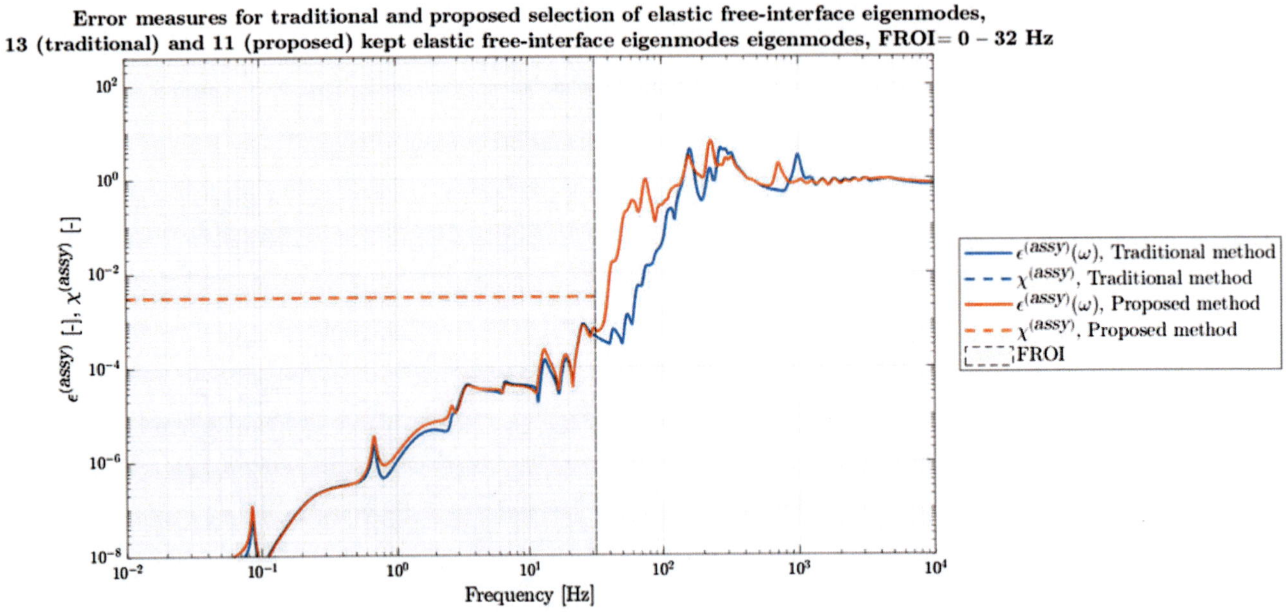

Fig. 1.6 Assembly receptance error measures $\epsilon^{(assy)}(\omega)$ and $\chi^{(assy)}$ obtained by using HH reduction bases where 13 elastic free-interface component eigenmodes are chosen using the traditional method and 11 using the proposed method. Note that the blue and red dashed lines practically coincide

not shown, can also be witnessed outside the FROI. This observation highlights that the proposed method focuses on approximating input–output behavior, but is also able to approximate eigenfrequencies with satisfactory accuracy.

The third use case shows how many eigenmodes should be selected using the traditional method such that the exact receptance error measure approximately equals the exact receptance error measure obtained using the sensitivity-based method in the second use case (where $n_{\mathcal{K}}^{(assy)} = 11$). As shown in Table 1.5, this is the case when 13 eigenmodes are selected using the traditional method. This situation is presented in Fig. 1.6, where the frequency-dependent error measures are comparable in the entire FROI. This use case thus shows that a smaller reduced model can be obtained with the proposed method, without increasing the reduction error compared to the traditional method.

1.5 Conclusions and Future Work

In this paper, a novel method for the selection of elastic free-interface substructure eigenmodes as used to reduce substructure models using the Hintz–Herting CMS method is proposed. The goal of this selection method is that the reduced assembly, obtained by coupling the reduced substructures, provides an accurate description of the frequency response functions at and between specific points of interest within a frequency range of interest. This goal is achieved by efficiently estimating assembly receptance reduction errors using sensitivities of the assembly receptance with respect to its substructure receptances and efficient descriptions of the substructure receptance error obtained using modal truncation error approximations. A leave-one-out approach is adopted to estimate the assembly receptance errors caused by the truncation of individual substructure eigenmodes. After condensing these estimated receptance errors using spatial and temporal norms, they are ranked by magnitude. The eigenmodes causing the highest approximate receptance errors are then selected for the set of kept eigenmodes as used in the substructure reduction bases. It has been shown that a selection of the eigenmodes obtained using the novel method results in accurately reduced assembly models. In fact, three use cases have been presented in which the sensitivity-based method outperforms the traditional selection method, which is solely based on cutoff frequencies. This observation highlights that the traditional method does not select an optimal set of retained eigenmodes. Therefore, using the sensitivity-based method for the selection of elastic free-interface eigenmodes, can provide the user with more accurate and/or efficient reduced assembly models than when the traditional selection method is used. Although, in this paper, the method has been applied to HH reduced substructure models, the method can be applied to other CMS methods using free-interface eigenmodes as well. Furthermore, it was explained that the selection of retained eigenmodes found by evaluating receptance errors is equivalent to evaluating mobility or inertance error matrices.

Despite the advantages shown here, it cannot be guaranteed that the proposed method will always outperform the traditional method, due to the approximative nature of the selection procedure. In fact, for some use cases based on the current demonstrator, the traditional method is found to perform (slightly) better. However, the used demonstrator is academic. For a (much) more complex engineering system such as a lithography machine, an airplane, or a ship, with more eigenmodes which act relatively locally, the sensitivity-based method is expected to only select local modes which are relatively important for specific assembly response behavior and truncate irrelevant (local) modes. It is expected that application of the sensitivity-based method to such more complex systems would show more clearly the benefit of the novel mode-selection method over the traditional selection method based on a cutoff frequency, certainly if the accuracy of only a limited number of FRFs (related to POIs) is important. This has to be investigated in the future. Furthermore, the novel selection method may be improved by adopting an iterative selection approach instead of the currently used leave-one-out approach. In such an iterative approach, the first eigenmode to be kept is selected as in the sensitivity-based method, but for each subsequently selected eigenmode it is taken into account that there is a set of eigenmodes already present in the reduced model. Although such an iterative method is computationally much more expensive, it might improve the reliability of the sensitivity-based method. Finally, instead of the first-order approximation of the assembly error used in this paper, higher-order approximations can be used. Even though this again comes at the cost of computation time, it could improve reliability of the proposed method as well.

References

1. Besselink, B., Tabak, U., Lutowska, A., Van De Wouw, N., Nijmeijer, H., Rixen, D.J., Hochstenbach, M.E., Schilders, W.H.: A comparison of model reduction techniques from structural dynamics, numerical mathematics and systems and control. J. Sound Vib. **332**(19), 4403–4422 (2013)
2. Craig, R.: Coupling of substructures for dynamic analyses – an overview. In: 41st Structures, Structural Dynamics, and Materials Conference and Exhibit (2000)
3. Craig, R., Bampton, M.: Coupling of substructures for dynamic analyses. AIAA J. **6**(7), 1313–1319 (1968)
4. De Klerk, D., Voormeeren, S.N.: Uncertainty propagation in experimental dynamic substructuring. In: Conference Proceedings of the Society for Experimental Mechanics Series, vol. 26 (2008)
5. De Klerk, D., Voormeeren, S.N.: Uncertainty propagation in experimental dynamic substructuring. Mech. Syst. Signal Proces. **24**, 106–118 (2010)
6. De Klerk, D., Rixen, D.J., De Jong, J.: The frequency based substructuring (FBS) method reformulated according to the dual domain decomposition method. In: Conference Proceedings of the Society for Experimental Mechanics Series, vol. 24 (2006)
7. de Kraker, B.: A Numerical-Experimental Approach in Structural Dynamics. Shaker Publishing, Maastricht (2004). ISBN 90-423-0259-3
8. Glover, K.: A tutorial on Hankel-norm approximation. In: From Data to Model, pp. 26–48 (1989)
9. Herting, D.N.: A general purpose, multi-stage, component modal synthesis method. Finite Elem. Anal. Des. **1**(2), 153–164 (1985)
10. Voormeeren, S.N.: Dynamic substructuring methodologies for integrated dynamic analysis of wind turbines. Ph.D. Thesis, Delft University of Technology (2012)

Chapter 2
Assessing the Quality of Real-Time Hybrid Simulation Tests with Deep Learning Models

Elif Ecem Bas and Mohamed A. Moustafa

Abstract Hybrid simulation (HS) is an advanced dynamic testing method that combines experimental testing and analytical modeling simultaneously to provide a better understanding of the structural systems as well as the structural elements while maintaining cost-effective solutions. A complex analytical substructure in the HS can be challenging, especially to conduct real-time HS (RTHS) tests due to the nature of numerical solution algorithms. Therefore, alternative methods, such as machine learning models are being explored to represent the analytical substructures of the RTHS tests. This study investigates the quality of the RHTS tests when a deep learning algorithm is used as a metamodel of the analytical substructure. A one-bay one-story concentrically braced frame (CBF) is selected to be used in RTHS tests where the frame is the analytical substructure, and the brace is tested experimentally. The compact HS laboratory at the University of Nevada, Reno, was used to run the RTHS experiments. Deep long short-term memory (LSTM) networks were selected to be trained as a metamodel using the Python environment to represent the dynamic behavior of the analytical substructure CBF. The pure analytical solution of the CBF under earthquake excitation is used as a training dataset of the metamodels. Several RTHS tests were performed. The quality of the test results was evaluated against the pure analytical solutions obtained from both the finite element model (FEM) and machine learning (ML) model.

Keywords Real-time hybrid simulation · Deep learning · Machine learning · Metamodel

2.1 Introduction

Hybrid simulation (HS) is a well-established testing method that couples experimental components with analytical models to evaluate structural elements and overall structural system behavior under realistic loading conditions (earthquake, wind, etc.). The first HS test was introduced in the early 1970s by Takanashi et al. [1] as "on-line testing," where the nonlinear differential equation was solved with updating the stiffness component of the spring-mass model from the structural experiment at each time step. Since then, there have been numerous studies to extend the variety of applications and the applicability of this technique to several hazards.

Each time step of the dynamic analysis in HS or RTHS is usually solved with direct integration methods for coupled experimental-computational model. Therefore, one of the main focuses of HS/RTHS studies is to develop numerical integration algorithms specialized on solving the equation of motion for substructured model for efficient and reliable experiment results (e.g., [2–4]). However, conducting real-time applications are still challenging, especially when the structure has complex computational substructure such as large degrees of freedoms in addition to numerical and/or experimental nonlinearities. It has been shown by Del Carpio et al. [5] that a careful sensitivity analysis is needed for large and complex structures to provide accurate and stable simulations. More recently, Bas and Moustafa [6] evaluated currently available direct integration algorithms for RTHS when computational models involve complex nonlinear behaviors. This study concludes that the current integration algorithms are not capable of conducting RTHS tests when some types of nonlinear behaviors are involved and experiments become more sensitive to hardware capabilities. On the other hand, another important focus of RTHS research studies is on accurate actuator control. There have been several efforts to compensate the delay and possible amplitude error in the response due to the nature of combined experimental substructure and the servo-hydraulic actuator [7–10].

E. E. Bas (✉) · M. A. Moustafa
Department of Civil and Environmental Engineering, College of Engineering, University of Nevada, Reno, NV, USA
e-mail: basel@nevada.unr.edu

M. S. Allen et al. (eds.), *Dynamic Substructures, Volume 4*, Conference Proceedings of the Society for Experimental Mechanics Series, https://doi.org/10.1007/978-3-030-75910-0_2

In order to tackle the current challenges and improve RTHS testing advancements, potential alternatives are being explored from various disciplines, and one of them is machine learning (ML) studies. ML has recently become a very popular area in earthquake engineering due to offering advantages such as providing computational efficiency, handling complex datasets, decision-making processes, and treating uncertainties [11]. Different ML algorithms have been recently used to predict structural dynamic responses [12–15]. Mucha [16] used artificial neural networks (ANNs) to represent the analytical substructure in HS to replace FEM of a bicycle frame and analyzed it under time-varying force. Moreover, a foundational work is conducted by Bas and Moustafa [17] to validate the communication developments and verify the Python-based deep learning ML metamodels. The study shows the communication validations successfully for RTHS tests with advanced deep learning ML models.

The main objective of this work is to investigate the quality of the ML models in RTHS tests. First this paper briefly introduces the HS setup and the system configuration where Python is used as computational environment. A one-bay one-story CBF is used to train and evaluate the metamodels under earthquake excitation. The columns and the beam considered to have material nonlinearity, where the experimental substructure is kept in linear range to make more careful comparisons. Deep LSTM networks are used to generate advanced ML models to represent analytical substructures of the RTHS tests. Several online RTHS experiments are conducted without test specimens and with test specimens. Moreover, the effect of using delay compensator is also investigated for the RTHS tests with and without specimen. In this study, the adaptive time series compensator (ATS) that is developed by Chae et al. [7] is used as a delay compensator. The test results are compared with the pure analytical metamodel predictions and pure analytical finite element model solutions to access the quality of the RTHS tests when advanced metamodels are used as computational substructures.

2.2 RTHS Setup and Configuration

The recently developed compact HS setup at the Large-Scale Structures Laboratory (LSSL) at the University of Nevada, Reno (UNR), is used for this study [18]. This setup is feasible for many studies including investigating computational challenges in HS/RTHS [6], developing new substructuring concepts [18], developing innovative approaches for computational substructures [17, 19, 20] as well as educational demonstrations.

The system consists of the following: (1) a small-scale load frame with a dynamic actuator run by an isolated hydraulic pump; (2) MTS STS controller with 4-channels with 2048 Hz clock speed; (3) real-time high performance Simulink machine (Speedgoat xPC Target); (4) Windows machine (Host PC) for MATLAB, OpenSees, Python, and the HS middleware OpenFresco [21]; and (5) SCRAMNet ring that provides shared memory locations for real-time communication.

In this setup, the finite-element (FE) model-based analytical substructure of the HS/RTHS can be either modeled in Simulink or OpenSees [19]. With the recent developments, Python can be also used as a computational environment to conduct RTHS with ML-based analytical substructures. The setup is capable of running both real-time and pseudo-dynamic (slow) HS tests. For the RTHS tests, the analytical models can be modeled in either Simulink or OpenSees. For the Simulink-based models, they can be directly compiled and run through xPC target. Other than this, the Host PC works as analytical substructure of the HS/RTHS tests. Furthermore, to synchronize the different time ranges between the simulation, integration, and controller, the predictor-corrector algorithm that is embedded in OpenFresco is used. Therefore, the rate of the test (i.e., constant velocity, pseudo-dynamic, real-time) can be controlled through OpenFresco.

For HS/RTHS experiments, a proper data communication between the analytical and experimental substructures is needed. OpenFresco middleware is developed to act as an interface software between FE models and the controllers and data acquisition systems in the laboratories to facilitate HS testing [22]. Moreover, it also allows to connect with different computational drivers to the HS/RTHS loop. Therefore, a novel communication scheme is developed to connect the Python environment to the HS/RTHS by using OpenFresco middleware [17]. The recent communication developments when Python-based ML models are used in the HS/RTHS is briefly explained in this section. It should be noted that there are two possible Python environment locations, namely, Local PC (Host PC) and cluster. The system can also use clusters (e.g., high-performance computers and supercomputers) as a computational platform with the current setting. In this section, only communication with the Local PC is explained in summary; however, the interested reader is referred to for more detailed information about the communication with clusters.

Figure 2.1 shows schematic communication details when Python-based metamodel substructures are located in Local PC (Host PC). The Host PC is also where the xPC connections are made to setup and SCRAMNetGT memory locations and initialize the simulations. The architecture of OpenFresco communication is called the "*client&middle-tier server.*" In this architecture, UDP/TCP channel is used for the OpenFresco and Python communication where Python is the client and OpenFresco is the middle-tier server. The UDP/TCP channel is opened at the server side (i.e., OpenFresco), and the

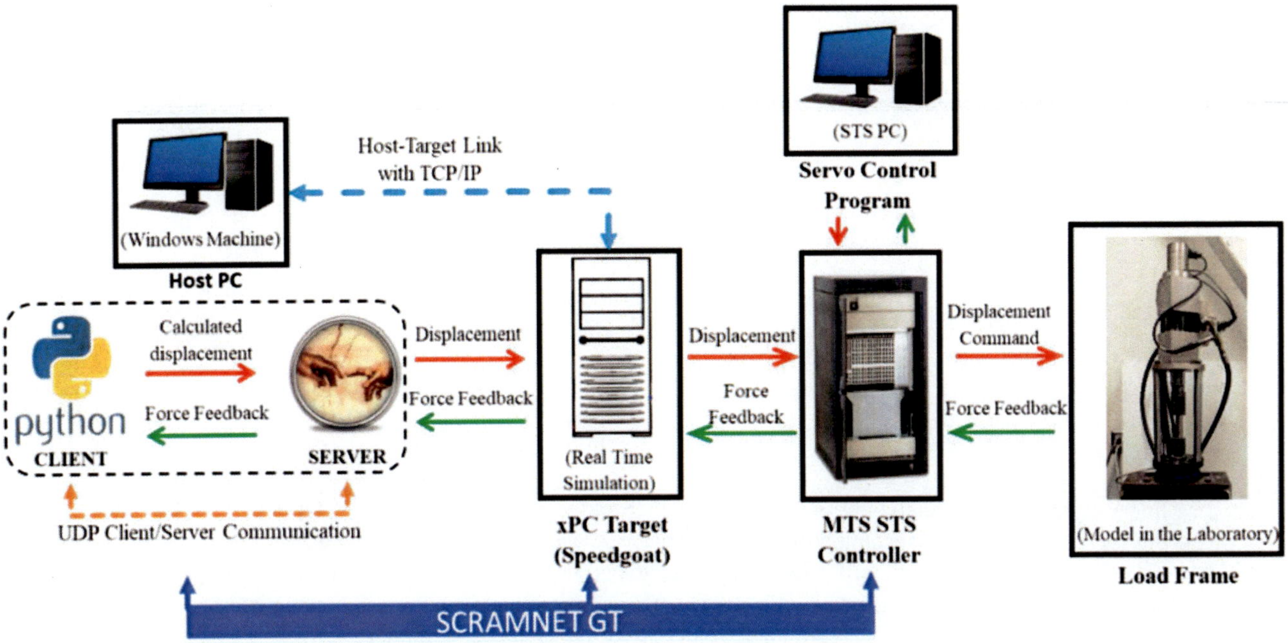

Fig. 2.1 HS/RTHS communication scheme for Python-based analytical substructure (metamodel) modeled in Local PC

simulation application site is set. Python has a module called "socket" that is used for the connection setup with either TCP or UDP protocols. Thus, Python (the client) connects the open provided server connection. In this communication development for HS/RTHS loop, the UDP protocol is used. On the contrary, OpenFresco also connects to the experimental site as a middle-tier server. In this study, the local experimental site is used. Moreover, the interface with the laboratory hardware is defined with the experimental control, where the data transfer between Host PC/controller is satisfied through xPC target. In this HS setup, the xPC target and the Host PC connection can be either provided with TCP/IP or SCRAMNetGT connection. The SCRAMNetGT secures more stable and faster data transfer over TCP/IP connection. Therefore, the SCRAMNetGT connection is used as the experimental control for this study.

In brief, the data transfer concept of the RTHS setting when Python-based ML model is used as a computational substructure is explained as follows. The computational substructure (metamodel) runs from Python, and the input displacements are calculated at each time step to be sent to the experimental substructure as displacement command. The force from the experimental substructure is measured from the specimen and fed back into the metamodel. As mentioned earlier, the time step synchronization between the controller, integration, and simulation is satisfied through the predictor-corrector algorithm that is defined in OpenFresco. For the computational substructures with metamodels, the integration time step becomes the prediction time step since there is no integration processed with metamodels. In this study, the prediction time step and the simulation time step are set to be 0.02 s for both the training stage and the RTHS tests.

2.3 Modeling Assumptions for Training Dataset

For the metamodels that are used to represent the analytical substructure behavior of the RTHS tests, deep LSTM networks are used. The training dataset is obtained from pure analytical responses of the overall structure. In this section, the FE model parameters that are used in the training of the metamodel and identifying the training dataset are introduced. The methodology for the deep learning algorithm is briefly explained with the hyperparameter calibrations of the metamodels in the next section.

The one-bay one-story concentrically braced frame (CBF) is selected to be trained under earthquake excitation and later used in the RTHS experiments. CBFs are suitable for substructuring where the braces experience complex behaviors that are hard to model numerically, therefore tested experimentally. Concurrently, the columns and beams of CBFs can be modeled with high accuracy that constructs the analytical substructure. In the HS setup presented in Sect. 2.2, a small-scale brace (i.e., experimental substructure) is combined with an analytical substructure that is a prototype steel frame at full-scale. In this study, the analytical substructure is considered to be modeled with material nonlinearity. The experimental substructure

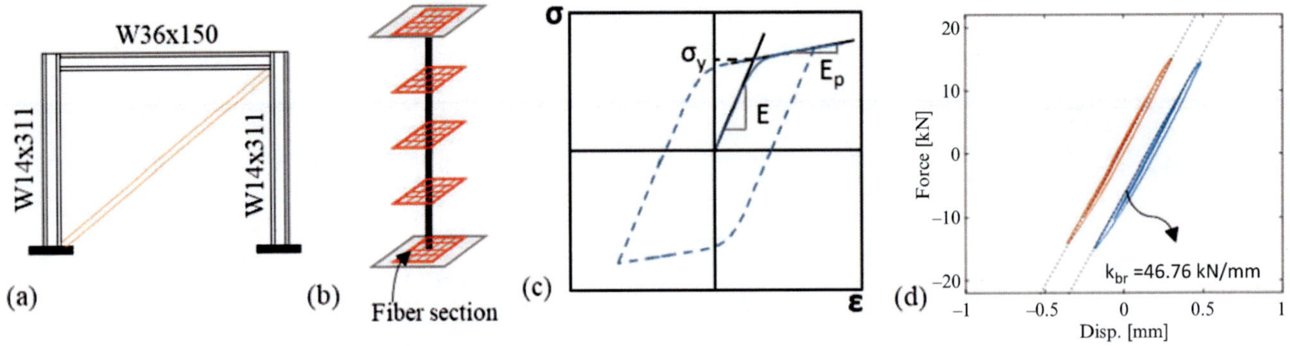

Fig. 2.2 (**a**) CBF model. (**b**) Nonlinear elements with distributed plasticity. (**c**) Stress-strain relationship of Steel 02 model. (**d**) Force-displacement hysteresis of the specimen under cyclic loading for linear range

is modeled as linear elastic in pure analytical models. Two RTHS cases are utilized, namely, without specimen and with the specimen. For the cases without a specimen, no physical brace is used; however, the actuator is free to move, and the actuator's displacement is multiplied with stiffness constant to generate force feedback to represent a hypothetical linear elastic test specimen. Furthermore, for the cases with the actual specimen, the specimen is kept in linear-elastic range in order to make a proper investigation on the performance of RTHS with deep learning models.

The pure analytical model of the overall CBF is modeled in OpenSees [23]. The columns (W14 × 311) and beam (W36 × 150) elements are modeled with fiber sections along with the distributed plasticity (Fig. 2.2a, b). The steel material model is defined with uniaxial Giuffre-Menegotto-Pinto steel material with isotropic strain hardening [24], which is Steel 02 in OpenSees (Fig. 2.2c). The yield stress of the material is selected to be 250 MPa, where the elastic modulus is 200 GPa. As mentioned earlier, the experimental substructure is modeled linear-elastic. In order to get the brace characteristics, two specimens are tested under increasing scale cyclic loading (Fig. 2.2d). The axial stiffness of the brace in linear elastic phase is obtained as 46.76 kN/mm. The geometric scale (S_L) is 25 for the small-scale brace, which comes in 419 kN/mm axial stiffness in prototype scale. Again, the experimental substructure is selected to be remain linear-elastic for both with and without specimen cases, to make a valid comparison for quality assessment.

In this study, to guarantee that the brace remains in linear elastic range, two different analytical models are considered with different mass assignment: 1.40 kN-s^2/mm and 1.75 kN-s^2/mm. The frame with lighter mass (Model 1) has the natural period of 0.22 s, whereas for the other frame (Model 2) it is calculated to be 0.25 s. The inherent damping of the structure is assumed to be 2% mass proportional Rayleigh damping.

The nonlinear time history analysis of the pure analytical CBFs is conducted for 1940 El-Centro earthquake. The implicit Newmark method (average acceleration) is selected as the integration algorithm with 0.001 s time steps. The duration of the earthquake is 31.2 s, but the analysis is carried out for 32 s. For the purpose of training of the metamodels, the dataset is resampled with 0.02 s time step, which lead 1600 data points for each response. The brace displacement time histories for Model 1 and Model 2 are given in Figs. 2.3a and 2.4a, respectively, where the yield displacement is marked with red dashed lines. It can be seen that for Model 1 the linear-elastic behavior is guaranteed for pure analytical response. Moreover, the global frame force-displacement hysteresis is also obtained for both models (Figs. 2.3b and 2.4b). From the responses, it can be concluded that Model 2 experiences higher nonlinearities, and the brace slightly exceeds the yield displacement at some points.

The training dataset inputs for the deep LSTM network are selected to be ground motion acceleration and the brace force time history, where the output (prediction) of the metamodel is the displacement of the brace, which is the input for the experimental substructure. It is well known the RTHS has a closed-loop system. Thus, the force of the brace (i.e., the experimental substructure) is dependent on the brace displacement. A high-level uncertainty is present due to this dependency, which can cause unstable predictions and lead erroneous test results. Furthermore, the load frame itself has its own uncertainty due to the nature of the servo-hydraulic system. Thus, a systematic bias is introduced to the brace force time history, and the training surface is expended with six more cases using offsets of ±5%, ±10%, and ±15% on the brace force. Finally, each input has 11,200 data points that are force input with bias, and the same ground motion repeated seven times.

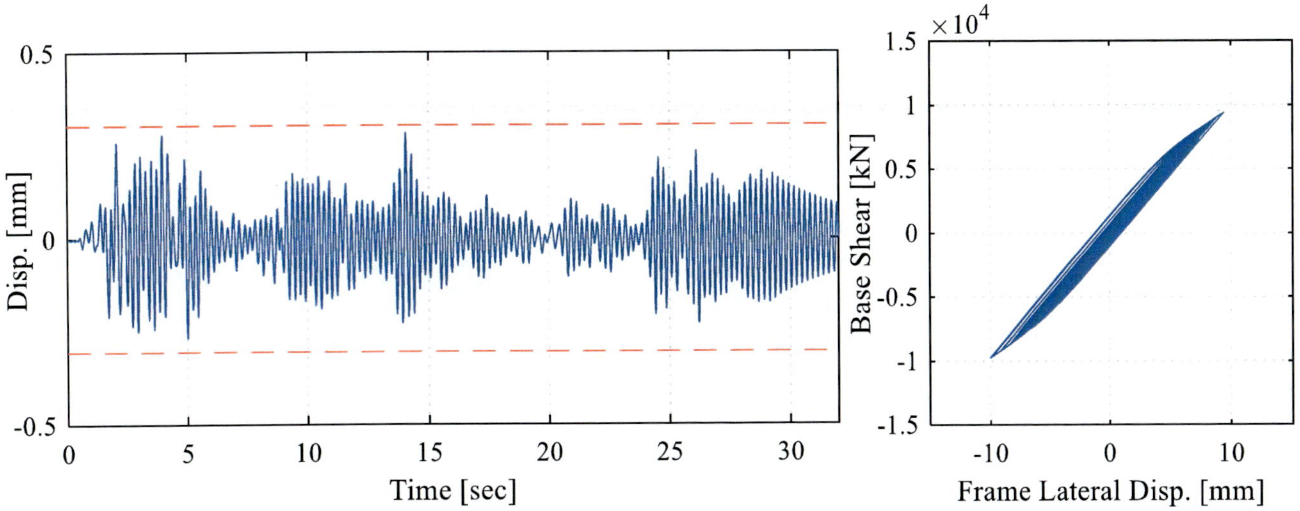

Fig. 2.3 (**a**) Brace displacement response time history. (**b**) Global force-displacement response for Model#1

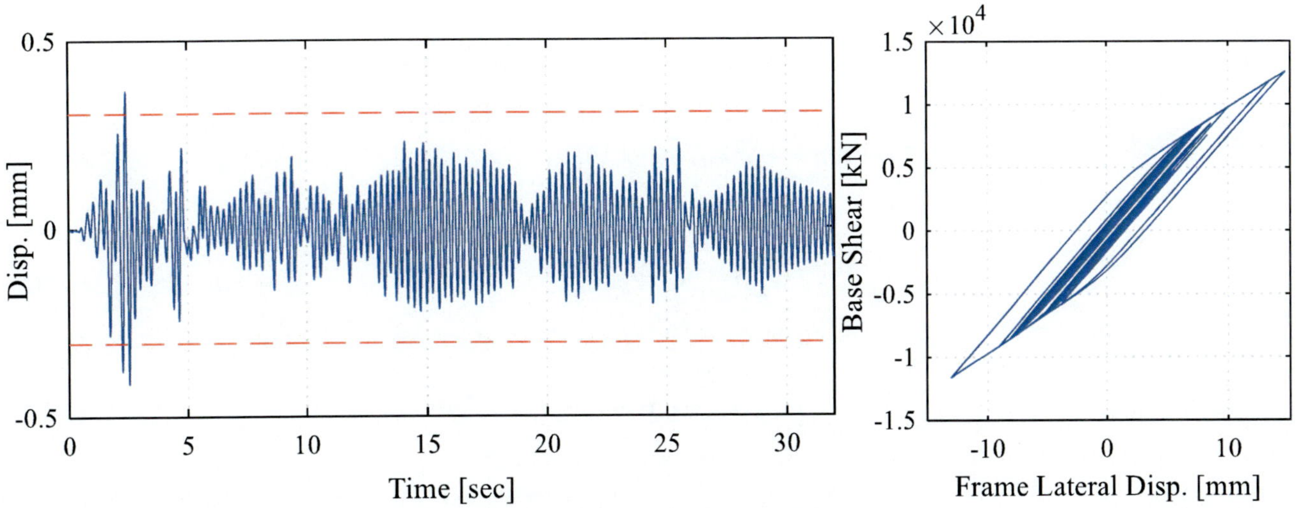

Fig. 2.4 (**a**) Brace displacement response time history. (**b**) Global force-displacement response for Model#2

2.4 Methodology for Deep LSTM Networks

As previously mentioned, deep LSTM networks are used to model the analytical substructure of the RTHS system. One of the most used ML algorithm to predict time series is Recurrent Neural Network (RNN). The advantage of RNN over the other ML algorithms is that it has a backward connection point; therefore, the layers can get an additional input that is the model output from the previous time step. However, this model has two main drawbacks, which are (i) having a limited short-term memory and (ii) having unstable gradients [25]. Thus, LSTM cells are developed to converge faster and detect the long-term dependencies of the datasets.

An LSTM network and a cell architecture are represented in Fig. 2.5. In every time step, the LSTM cell receives two additional inputs other than ($\mathbf{x}_{(t)}$), which are to represent short-term states ($\mathbf{h}_{(t-1)}$) and long-term states ($\mathbf{c}_{(t-1)}$). The previous time step output vector ($\mathbf{h}_{(t-1)}$) and current time step inputs ($\mathbf{x}_{(\mathrm{t})}$) are fed into four different fully connected layers. In a regular RNN cell, there is only $\mathbf{g}_{(\mathrm{t})}$ layer, which is the weighted sum of the inputs with an activation function of tanh. The other three gate controllers are the ones that help to control memory information for longer sequences and use the logistic function as an activation function (0 or 1). The long-term state's unnecessary parts are deleted at the forget gate (output of $\mathbf{f}_{(t)}$). The input gate controls which parts of $\mathbf{g}_{(\mathrm{t})}$ should be added in the long-term state. Moreover, finally, the parts of the long-term state should be the output to both $\mathbf{h}_{(\mathrm{t})}$, and the output gate ($\mathbf{o}_{(t)}$) manages $\mathbf{y}_{(t)}$. The equations for the LSTM cell computations are briefly given in Eqs. (2.1), (2.2), (2.3), (2.4), (2.5), and (2.6). In the equations, $\boldsymbol{W}_{xi}$, $\boldsymbol{W}_{xf}$, $\boldsymbol{W}_{xf}$, $\boldsymbol{W}_{xf}$, are the

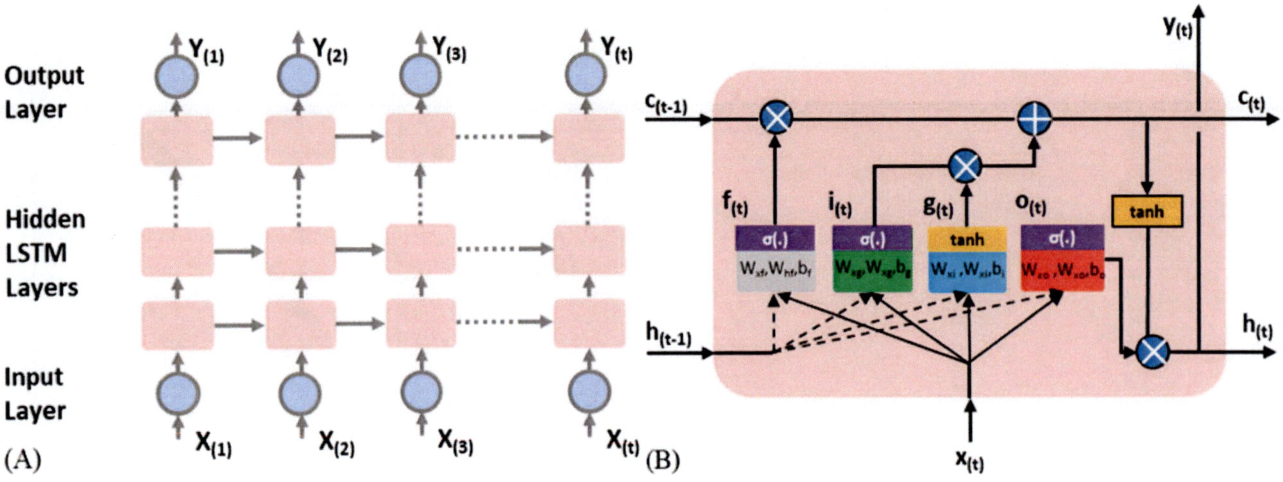

Fig. 2.5 (**a**) A deep LSTM network through time. (**b**) LSTM cell architecture

weighted matrices for the input vector, and $\boldsymbol{W}_{hi}$, $\boldsymbol{W}_{hf}$, $\boldsymbol{W}_{hf}$, $\boldsymbol{W}_{hf}$ are the weighted matrices of the previous short-term state vector $\mathbf{h}_{(t-1)}$ for each layer. Moreover, every layer has the bias term which are $\boldsymbol{b}_i$, $\boldsymbol{b}_f$, $\boldsymbol{b}_o$ and $\boldsymbol{b}_g$.

$$\boldsymbol{i}_{(t)} = \sigma\left(\boldsymbol{W}_{xi}{}^{\mathrm{T}}\boldsymbol{x}_{(t)} + \boldsymbol{W}_{hi}{}^{\mathrm{T}}\boldsymbol{h}_{(t-1)} + \boldsymbol{b}_i\right) \tag{2.1}$$

$$\boldsymbol{f}_{(t)} = \sigma\left(\boldsymbol{W}_{xf}{}^{\mathrm{T}}\boldsymbol{x}_{(t)} + \boldsymbol{W}_{hf}{}^{\mathrm{T}}\boldsymbol{h}_{(t-1)} + \boldsymbol{b}_f\right) \tag{2.2}$$

$$\boldsymbol{o}_{(t)} = \sigma\left(\boldsymbol{W}_{xo}{}^{\mathrm{T}}\boldsymbol{x}_{(t)} + \boldsymbol{W}_{ho}{}^{\mathrm{T}}\boldsymbol{h}_{(t-1)} + \boldsymbol{b}_o\right) \tag{2.3}$$

$$\boldsymbol{g}_{(t)} = \tanh\left(\boldsymbol{W}_{xg}{}^{\mathrm{T}}\boldsymbol{x}_{(t)} + \boldsymbol{W}_{hg}{}^{\mathrm{T}}\boldsymbol{h}_{(t-1)} + \boldsymbol{b}_g\right) \tag{2.4}$$

$$\boldsymbol{c}_{(t)} = \boldsymbol{f}_{(t)} \otimes \boldsymbol{c}_{(t-1)} + \boldsymbol{i}_{(t)} \otimes \boldsymbol{g}_{(t)} \tag{2.5}$$

$$\boldsymbol{y}_{(t)} = \boldsymbol{h}_{(t)} = \boldsymbol{o}_{(t)} \otimes \tanh\left(\boldsymbol{c}_{(t)}\right) \tag{2.6}$$

The training of the deep LSTM model is done in Python environment using Tensorflow 2.0 [26]. The LSTM model input sequences require three-dimensional (3D) arrays, which are batch size (number of samples of the dataset), lookback (previous time steps), and size of the input dimension (number of the features) [25]. The hyperparameters of the LSTM network are tuned by feeding the inputs and the output datasets to the model. The optimizer of the training is selected to be the Adam (Adaptive Moment Estimator) optimizer with a learning rate of 0.001 [27]. The number of epochs is set to be 10^3. Moreover, the model is trained to minimize the cost function of mean squared error (MSE). From the previous experience [17], two deep learning models are trained to be used with 10 and 15 lookbacks, where each model has an input layer, four LSTM layers with 30 units in each layer, and one dense layer, which is a fully connection layer that outputs the prediction.

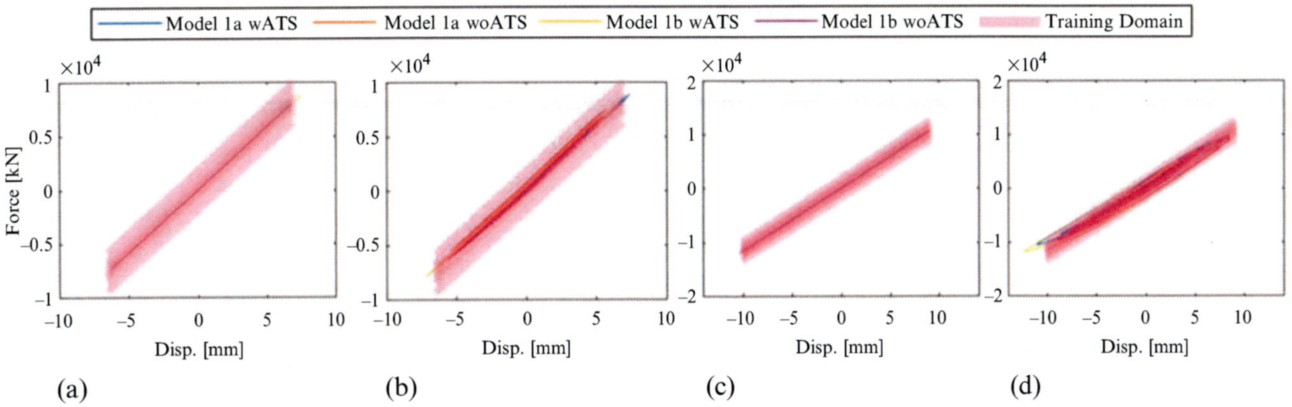

Fig. 2.6 Brace axial force-displacement response for Model 1 from the RTSH for (**a**) without specimen case; (**b**) with specimen case; for Model 2 for Model 2 from the RTSH for (**c**) without specimen case; (**d**) with specimen case

2.5 Online RTHS Test Results

In this section, the RTHS test results for the two CBF models, i.e., Model #1 and Model #2, where the trained deep LSTM networks were used as a computational substructure, were evaluated. Moreover, each model has two varieties, where 15 and 20 lookbacks are considered. Henceforth, the models with 15 lookbacks are referred to as Model 1a and Model 2a, and the ones with 20 lookback are referred to as Model 1b and Model 2b. Two sets of RTHS tests were considered to evaluate the quality of the ML models. The first set of RHTS tests were conducted with a hypothetical specimen where no actual specimen was used, but the actuator was free to move, and the force feedback was represented by multiplying the actual actuator displacement by a constant stiffness value of the stiffness of the specimen. In the second set of tests, actual specimens were used where the selected ground motion and frame characteristics were supposed to keep the brace within the linear elastic range. Moreover, all RTHS tests were conducted twice to evaluate the delay compensator (i.e., with and without ATS compensator). The last variable that was considered for this study was the lookback dimension of the force input where either the first or last dimension was updated with the actuator feedback. Thus, 32 different RTHS tests were conducted and selected test results are shown here.

First, the experimental substructure's axial force-displacement relationship for each model and each RTHS test case (i.e., with specimen and without specimen) is presented. The test results presented here is where only the first term of the force input through the lookback direction is updated. The axial force-displacement hysteresis of the brace for Model 1a, Model 1b, Model 2a, and Model 2b are shown in Fig. 2.6. As explained earlier, the training domain for the force feedback was extended with a systematic bias due to force-displacement dependencies and initial load frame feedback. The RTHS test results for with a hypothetical specimen use is shown in Fig. 2.6a and c. It can be seen from the figure that the brace behaviors for each case are obtained as expected for the online RTHS without the actual specimen. Moreover, the test results for the real specimen cases are shown in Fig. 2.6b and d. It can be seen from Fig. 2.6b that the actual brace remains linear elastic through the RTHS tests for Model 1. The actual brace remains linear elastic through the RTHS tests, which provides reliable test results when deep learning models are considered as computational substructures. As mentioned earlier, it is observed that during the training phase, Model 2 slightly exceeding the yield displacement. It can be seen from Fig. 2.6b that the actual specimen can have some small loops, which is not a severe nonlinearity. Moreover, the specimen is used repeatedly, which can be considered linear-elastic in this case. For both models, increasing the training domain is worked well, since the brace axial force-displacement relationships fit well in the training domain. Overall, no stability problem occurred and the tests are successfully conducted when an actual specimen is considered. Finally, it is observed that using a delay compensator did not affect the test results significantly for both models and each case.

Force and displacement time histories for the experimental substructure responses are compared with the training force input and output displacement datasets. This evaluation is done for Model 1a, where no delay compensator is used and last input dimension of the force input is updated with the force feedback from the experimental setup. Figure 2.7 shows the brace displacement time history and brace force time history for Model 1a where a hypothetical brace is used. It can be seen from the displacement time histories, the prediction error starts at the early stages of the RTHS test, and these errors are accumulated with the hardware setup and led wrong experimental results. The force feedback obtained from the experimental setup is shown in Fig. 2.7b. As expected, the prediction error, which is also the input command for the actuator, is reflected to the force feedback. The RTHS tests are unsuccessful, but no stability issues are observed.

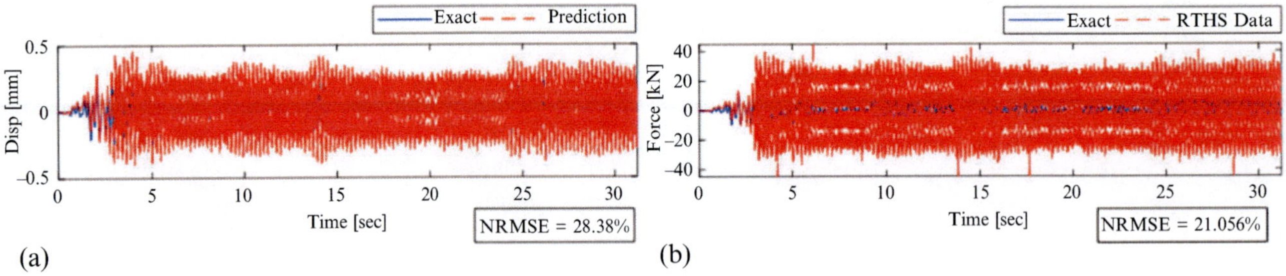

Fig. 2.7 Model 1a without specimen case: (**a**) brace displacement time history prediction comparison with exact solution, and (**b**) brace force feedback time history prediction comparison with exact solution

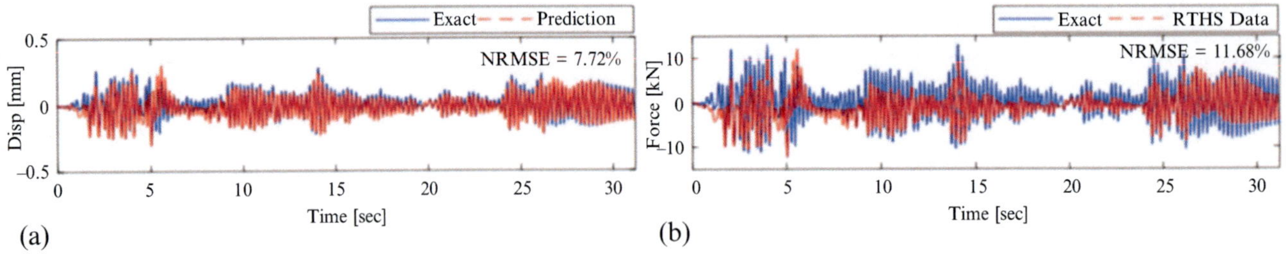

Fig. 2.8 Model 1a with specimen case: (**a**) brace displacement time history prediction comparison with exact solution, and (**b**) brace force feedback time history prediction comparison with exact solution

Table 2.1 NRMSE [%] for brace displacement predictions where "first" force input dimension is replaced in RTHS tests

		Model 1		Model 2	
		Model 1a	Model 1b	Model 2a	Model 2b
Virtual Hybrid Simulation		1.05	1.02	1.02	1.13
w/ATS	w/o Specimen	1.06	1.03	1.04	1.12
	w/ Specimen	1.03	1.03	1.06	1.10
w/o ATS	w/o Specimen	0.97	1.02	1.02	1.12
	w/ Specimen	0.93	1.02	1.02	1.11

Table 2.2 NRMSE [%] for brace displacement predictions where "last" force input dimension is replaced in RTHS tests

		Model 1		Model 2	
		Model 1a	Model 1b	Model 2a	Model 2b
Virtual Hybrid Simulation		9.33	5.46	10.48	13.02
w/ATS	w/o Specimen	29.78	20.85	26.32	27.16
	w/ Specimen	20.02	17.39	12.07	12.94
w/o ATS	w/o Specimen	28.37	2.04	27.65	28.42
	w/ Specimen	7.72	5.01	6.54	23.88

Figure 2.8 shows the test results for the same case where a real specimen is used this time. It can be seen from Fig. 2.8a that, when an actual specimen is present within the HS loop, the prediction error in the early stages of the model did not accumulate which in return resulted more reliable tests results. Although the predictions are not that accurate, it gives satisfactory enough predictions which are not noisy as in the previous case. Moreover, the system dynamics are also represented well. Figure 2.8b shows the brace force feedback obtained in return to the predicted brace displacement. As it is seen from the force time history, the force feedback is appropriate and relevant with the displacement input. Therefore, including a real specimen within the system improved the performance of the RTHS test results with the ML computational substructures.

Finally, the normalized root mean square errors (NRMSEs) are calculated for all conducted RTHS cases and listed in Tables 2.1 and 2.2 focus on the NRMS errors for the brace displacement prediction for the last input dimension is updated with the experimental feedback. When this is considered, the use of ATS makes models more sensitive to the error in the predictions. Therefore, the use of delay compensator with the ML models can be eliminated.

Table 2.2. In these tables, the brace displacement predictions are compared with the training displacement data. Before conducting the tests, pure analytical hybrid model representations are tested, where at every time step, the displacement predicted by the deep learning model and the input force is calculated by multiplying with the stiffness constant. It can be seen from the results that the models have more errors when the last lookback dimension is updated for the pure analytical response. However, for Model 1, this error is reduced for the cases with an actual specimen where no delay compensator is used. On the other hand, the error is more pronounced for Model 2 with the last dimension update. This error accumulated when the online RTHS tests are conducted. The specimen use slightly helped for stabilizing and reducing the error. For all cases where the first parameter of the force input's lookback dimension is changed, it performs well without any noise issues in the predictions.

Table 2.2 focuses on the NRMS errors for the brace displacement prediction for the last input dimension is updated with the experimental feedback. When this is considered, the use of ATS makes models more sensitive to the error in the predictions. Therefore, the use of delay compensator with the ML models can be eliminated.

2.6 Summary and Concluding Remarks

This study focuses on providing information on the quality of the RTHS results when ML models are used to represent computational substructure behavior under dynamic excitation. Thus, two FEM models of one-bay one-story CBF are developed to obtain a training dataset for the ML models. Deep LSTM networks are used to develop the metamodels of the computational substructure's response under earthquake excitation. Two metamodels are developed for each model with 15 and 20 lookbacks. Overall, a total of 32 RTHS tests are conducted. The RTHS tests mainly focus on two main aspects, which are with and without an actual specimen. When the actual specimen is used in the online RTHS tests, the specimen aims to ensure linear-elastic behavior. In each case, the force feedback from the actuator is either updated in first dimension of the force input or last dimension.

The experimental substructure responses of the brace hysteresis are represented for the cases where the force input's first dimension is updated. For Model 1 and Model 2, the deep learning models give satisfactory results. The responses fit within the training domain, which shows that generating a larger training domain for the uncertainties that could occur due to the brace force-displacement and load frame is covered well.

Moreover, the force and displacement time histories are also evaluated. It is shown that once the RTHS model does not have an actual specimen, the error accumulates and gives wrong predictions. These wrong predictions lead to wrong experimental results. However, this error accumulation is resolved once an actual specimen is placed to the RTHS system. Although the model is still wrong, it reduces the noise in the experimental system.

Overall, the RTHS tests performed well when deep learning metamodels are used to represent the analytical substructure dynamic behavior and no stability problem observed. However, for the models where the first dimension of the force input is updated, performed better. Finally, with the use of ML models within the HS as analytical substructure response, the use of delay compensator can be eliminated, since the models can be trained with this delay beforehand. However, more investigations are needed to make more reliable recommendations.

References

1. Takanashi, K., Udagawa, K., Seki, M., Okada, T., Hisashi, T.: Non-linear earthquake response analysis of structures by a computer-actuator on-line system. Bull. Earthq. Resist. Struct. Res. Cent. (No. 8). Japan Inst. Ind. Sci. Univ. Tokyo (1975)
2. Chang, S.-Y.: A family of noniterative integration methods with desired numerical dissipation. Int. J. Numer. Methods Eng. (July), 62–86 (2014). https://doi.org/10.1002/nme
3. Chen, C., Ricles, J.M., Marullo, T.M., Mercan: Real-time hybrid testing using the unconditionally stable explicit CR integration algorithm. Earthq. Eng. Struct. Dyn. **38**, 23–44 (2009). https://doi.org/10.1002/eqe.838
4. Bonelli, A., Bursi, O.S.: Predictor-corrector procedures for pseudo-dynamic tests. Eng. Comput. (Swansea, Wales). **22**(7), 783–834 (2005). https://doi.org/10.1108/02644400510619530
5. Del Carpio, M., Hashemi, M.J., Mosqueda, G.: Evaluation of integration methods for hybrid simulation of complex structural systems through collapse. Earthq. Eng. Eng. Vib. **16**(4), 745–759 (2017). https://doi.org/10.1007/s11803-017-0411-z
6. Bas, E.E., Moustafa, M.A.: Performance and Limitations of Real-Time Hybrid Simulation with Nonlinear Computational Substructures. Exp. Tech. (2020). https://doi.org/10.1007/s40799-020-00385-6
7. Chae, Y., Kazemibidokhti, K., Ricles, J.M.: Adaptive time series compansator for delay compensation of servo-hydraulic actuator systems for real-time hybrid simulation. Earthq. Eng. Struct. Dyn. **42**, 1697–1715 (2013). https://doi.org/10.1002/eqe.2294

8. Zhao, J., French, C., Shield, C., Posbergh, T.: Considerations for the development of real-time dynamic testing using servo-hydraulic actuation. Earthq. Eng. Struct. Dyn. **32**(11), 1773–1794 (2003). https://doi.org/10.1002/eqe.301
9. Carrion, J.E., Spencer, B.F.: Model-based strategies for real-time hybrid testing. No. December, p. 211 (2007)
10. Wallace, M.I., Wagg, D.J., Neild, S.A.: An adaptive polynomial based forward prediction algorithm for multi-actuator Real-time dynamic substructuring. Proc. R. Soc. A Math. Phys. Eng. Sci. **461**(2064), 3807–3826 (2005). https://doi.org/10.1098/rspa.2005.1532
11. Xie, Y., Eeri, M., Sichani, M.E., Padgett, J.E., Eeri, M., and Eeri, M.: The promise of implementing machine learning in earthquake engineering: A state-of-the-art review. (2020), https://doi.org/10.1177/8755293020919419
12. Zhang, R., Liu, Y., Sun, H.: Physics-informed multi-LSTM networks for metamodeling of nonlinear structures. Comput. Methods Appl. Mech. Eng. **369**, 1–21 (2020)
13. Raissi, M., Perdikaris, P., Karniadakis, G.E.: Physics-informed neural networks: A deep learning framework for solving forward and inverse problems involving nonlinear partial differential equations. J. Comput. Phys. **378**, 686–707 (2019). https://doi.org/10.1016/j.jcp.2018.10.045
14. Bas, E.E., Aslangil, D., Moustafa, M.A.: Predicting nonlinear seismic response of structural braces using machine learning. (2020)
15. Zhang, R., Liu, Y., Sun, H.: Physics-guided Convolutional Neural Network (PhyCNN) for data-driven seismic response modeling, pp. 1–24 (2019)
16. Mucha, W.: Application of artificial neural networks in hybrid simulation. Appl. Sci. **9**(21) (2019). https://doi.org/10.3390/app9214495
17. Bas, E.E., Moustafa, M.A.: Communication Development and Verification for Python-Based Machine Learning Models for Real-Time Hybrid Simulation. Front. Built Environ. https://doi.org/10.3389/fbuil.2020.574965
18. Bas, E.E., Moustafa, M.A., Pekcan, G.: Compact hybrid simulation system: Validation and applications for braced frames seismic testing. J. Earthq. Eng. **00**(00), 1–30 (2020). https://doi.org/10.1080/13632469.2020.1733138
19. Bas, E.E., Moustafa, M.A., Feil-Seifer, D., Blankenburg, J.: Using a machine learning approach for computational substructure in real-time hybrid simulation. In: Conference Proceedings of the Society for Experimental Mechanics Series, 2020, no. Dynamic Substructures, vol. 4, https://doi.org/10.1007/978-3-319-32669-6
20. Bas, E.E., Moustafa, M.A.: Real-Time Hybrid Simulation with Deep Learning Computational Substructures: System Validation Using Linear Specimens. Mach. Learn. Knowl. Extr. **2**(4), 469–489 (2020). https://doi.org/10.3390/make2040026
21. Schellenberg, A.H., Kim, H.K., Mahin, S.A.: OpenFresco. Universtiy of California, Berkeley (2009)
22. Schellenberg, A., Kim, H.K., Fenves, G.L., Mahin, S.A.: TR-2009- [ID] OpenFresco framework for hybrid simulation: Simulation finite element adapter experimental control example. Eng. Simul. (2009)
23. McKenna, F., Fenves, G.L., Scott, M.H.: Open System for Earthquake Engineering Simulation. Pacific Earthq. Eng. Res. Cent. (2000)
24. Filippou, F.C., Popov, E.P., Bertero, V.V.: Effects of bond deteroriation on hysteretic behavior of reinforced concrete joints, University of California Berkeley, Berkeley. Report No. UCB/EERC-83/19 (1983)
25. Géron, A.: Hands-on Machine Learning with Scikit-Learn and TensorFlow. O'Reilly Media, Inc (2017)
26. Abadi, M. et al.: TensorFlow: Large-Scale Machine Learning on Heterogeneous Distributed Systems, Softw. available from tensorflow.org, arXiv:1603.04467 (2015)
27. Kingma, D.P., Ba, J.L.: Adam: A method for stochastic optimization. In: 3rd International Conference on Learning Representations ICLR 2015 – Conference Track Proceedings, pp. 1–15 (2015)

Chapter 3
Dynamic Substructuring Using a Combination of Softening and Hardening Connecting Elements

Jacopo Brunetti, Walter D'Ambrogio, Annalisa Fregolent, and Francesco Latini

Abstract The dynamic analysis of complex engineering structures represents a challenging task, since the reliability of the results significantly depends on the accuracy of the model. In general, linearized models represent a valid approximation, but in some cases, it is necessary to include also the most significant nonlinearities to obtain reliable results. In the present work, the case in which two beams are jointed together through softening and hardening connecting elements is analyzed. It is possible to account for their presence by modeling them as nonlinear substructures, and the connected subsystems are instead modeled as linear substructures. A Nonlinear Coupling Procedure (NLCP) is defined in the modal domain to analyze the dynamics of these systems. The iterative procedure has been modified with respect to the one used in previous works by selecting a different initial guess and by maintaining the energy of the system constant at each iteration. The theory of Nonlinear Normal Modes (NNMs) is used to account for the presence of the nonlinear connections in the coupled assembly. The NLCP is employed to analyze the effects of modal truncation on the mode shapes and on the resonance frequency.

Keywords Dynamic nonlinear substructuring · Nonlinear normal modes · Continuous systems · Modal coupling · Softening connection

3.1 Introduction

Most of the mechanical systems currently employed in industrial applications are composed of multiple components connected together. The evaluation of the dynamics of these assemblies might be challenging if they are considered as a single complex system. Over the years, dynamic substructuring techniques have been developed to tackle these problems in the physical, modal, and frequency domains [1]. Each component is considered as a substructure, and the behavior of the complete system can be achieved with a lower computational effort from the known behavior of its components. Using these techniques, it is also possible to combine experimental and numerical models [2, 3], which might be convenient to analyze how a new component affects the dynamics of an existing assembly. However, when using these techniques, the systems are considered as linear. This represents a limit in all those cases in which the behavior of the involved substructures is known to be nonlinear.

During the last years, many attempts have been made in order to employ the advantages of the substructuring techniques to analyze nonlinear systems. Kalaycioğlu and Özgüven propose an approach in the frequency domain to account for the presence of nonlinearities and perform coupling and decoupling procedures. In [4, 5] multiple substructures with localized nonlinearities are coupled together, while in [6] the decoupling problem is addressed. Note that the approach used in the coupling problem derives from the structural modification technique developed by Özgüven [7], and the presence of the nonlinearity is taken into account by using Describing Functions (DFs) [8]. Kuether et al. [9] and Mahdiabadi et al. [10] develop a non-intrusive technique to account for geometric nonlinearities and obtain a Nonlinear Reduced Order Model (NLROM) by using the Implicit Condensation and Expansion (ICE) method proposed by Hollkamp [11]. Kuether and Allen

J. Brunetti · W. D'Ambrogio
Dipartimento di Ingegneria Industriale e dell'Informazione e di Economia, Università dell'Aquila, L'Aquila, Italy
e-mail: jacopo.brunetti@univaq.it; walter.dambrogio@univaq.it

A. Fregolent · F. Latini (✉)
Dipartimento di Ingegneria Meccanica e Aerospaziale, Università di Roma La Sapienza, Rome, Italy
e-mail: annalisa.fregolent@uniroma1.it; francesco.latini@uniroma1.it

M. S. Allen et al. (eds.), *Dynamic Substructures, Volume 4*, Conference Proceedings of the Society for Experimental Mechanics Series, https://doi.org/10.1007/978-3-030-75910-0_3

[12] propose a substructuring technique in the modal domain by reducing each substructure using a set of Nonlinear Normal Modes (NNMs) [13, 14].

The present work aims at analyzing the effects of nonlinear connections on the dynamics of an assembly under the assumption that the coupled substructures can be described by linear equations. This hypothesis is valid when the connecting element is significantly more nonlinear than the substructures to be coupled in the considered working range, as it happens with bolted joints [15], wire rope isolators [16], and blades connected to a disk [17]. Thus, the nonlinearity is isolated and considered as an additional substructure to be included in the substructuring procedure. This method is inspired by the work of Kuether and Allen [12] who proposed an energetic approach to obtain the NNMs of the complete system by performing a substructuring procedure in the modal domain. The method proposed in this work, called Nonlinear Coupling Procedure (NLCP), is performed to couple linear substructures connected through nonlinear elements. The results obtained by applying the NLCP are compared to those provided using the Harmonic Balance method (HB) [18] on the whole system. The NLCP was already used to couple lumped parameter models connected by hardening and softening nonlinear springs [19, 20]. An experimental setup composed of two beams connected through two nonlinear connecting elements was also built in order to compare the results of the substructuring procedure with the data measured on the real system [21].

Significant enhancements have been performed with respect to the procedure used in previous works [19, 20]: on one hand, there is a better control of the total energy at each step of the procedure by adjusting the energy distribution among the substructures; on the other hand, the choice of the initial guess at each energy level is modified in order to increase the convergence rate. Here, the method is applied to couple two beams through two connecting elements, one being softening and the other being hardening. The results of the NLCP are compared to those provided by the HB method to evaluate its accuracy, and the effects of modal truncation are investigated.

3.2 Nonlinear Coupling Procedure

Linear substructuring techniques are commonly used to couple substructures in physical, modal, and frequency domain [1]. Several attempts have been made over the years to extend those methods to nonlinear systems, both in frequency [5, 6] and in modal domain [12, 22]. These procedures are based on the assumption that the nonlinear internal forces are harmonic at the same frequency of the excitation, thus higher harmonic effects are not detected. The technique described in this section is valid under the same assumption, and it is used to couple linear substructures presenting nonlinear connections, as the system shown in Fig. 3.1.

When performing the coupling procedure in the modal domain, a modal reduction basis for each substructure is considered. In the present case, Linear Normal Modes (LNMs) are used for the linear substructures while NNMs are used for the nonlinear substructures. Thus, instead of computing the NNMs of the whole system, they are computed only for the substructures representing the nonlinear connections. This leads to a significant decrease in the computational time, especially if dealing with complex systems.

The NLCP is an iterative technique that aims at obtaining the resonance frequency and the mode shapes of the coupled system as function of the total energy stored in it. The dependency of NNMs as function of energy requires the definition of an array $\{e\} = \{e_1 \cdots e_k \cdots\}^T$ that collects the e_k energy levels at which the coupling is performed. Significant improvements have been made with respect to the procedure used in [19, 20] with the purpose of controlling the energy levels at which each NNM is evaluated [23]. In doing so, it is possible to retrace the NNM frequency–energy curve at pre-set energy levels and build an iterative continuation technique where the solutions at e_{k-2} and e_{k-1} are used to estimate the initial guess for the energy level e_k, as shown in Fig. 3.2.

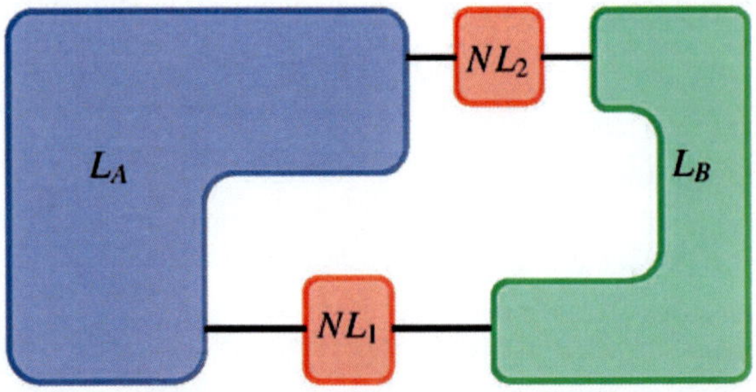

Fig. 3.1 Example of two substructures connected through two nonlinear connection

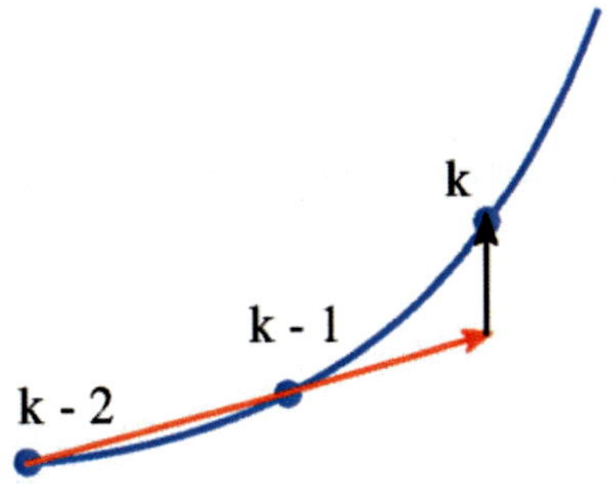

Fig. 3.2 Continuation Procedure: the blue solid line is the NNM to be identified, the red arrow indicates the prediction step, the black one refers to the correction step, and the blue dots are the identified points of the NNM

3.2.1 Nonlinear Coupling in Modal Domain

The Nonlinear Normal Modes of the connecting element depend on its energy level, thus it is necessary to derive the energy distribution among the substructures for a given energy amount of the coupled structure. To this aim, an iterative algorithm is used to find the energy of each substructure in the coupled configuration and to compare it to the energy found at the previous iteration, until convergence is achieved. The equations of motion describing the nonlinear system in the physical domain are

$$\begin{cases} [M]\{\ddot{x}\} + [K]\{x\} + \{f_{nl}(\{x\})\} = \{g\} \\ [B]\{x\} = \{0\} \\ [L]^T\{g\} = \{0\} \end{cases} \tag{3.1}$$

where

$$[M] = \text{diag}\left([M]^A, [M]^{CE}, [M]^B\right), \quad [K] = \text{diag}\left([K]^A, [K]^{CE}, [K]^B\right), \tag{3.2}$$

and $\{f_{nl}(\{x\})\}$ is the vector representing the nonlinear term, $\{g\}$ is the vector of connecting forces, $[B]$ is a signed Boolean matrix, and $[L] = \text{null}([B])$.

Supposing to know how the energy is distributed among the substructures for the first iteration, it is possible to pass from physical to modal coordinates through the reduction matrix $[R]$:

$$\{x\} \approx [R]_{k,i}\{y\} \tag{3.3}$$

with

$$[R]_{k,i} = \text{diag}\left([\Phi]^A, \left[\Psi(E^{CE}_{k,i})\right]^{CE}, [\Phi]^B\right) \tag{3.4}$$

where i refers to the iteration and k to the energy level. The matrices $[\Phi]^A$, $[\Phi]^B$ contain the set of eigenvectors of the linear substructures whereas $[\Psi]^{CE}$ contains the set of nonlinear modes computed at energy $E^{CE}_{k,i}$ for the nonlinear connecting element.

Therefore, Eq. (3.1) becomes

$$[I]\{\ddot{y}\} + [\Lambda]_{k,i}\{y\} = [R]^T_{k,i}\{g\} \tag{3.5}$$

with

$$[\Lambda]_{k,i} = \text{diag}\left([\Lambda]^A, \left[\Lambda(E^{CE}_{k,i})\right]^{CE}, [\Lambda]^B\right)s \tag{3.6}$$

where $[\Lambda]^A$ and $[\Lambda]^B$ are the matrices of eigenvalues of the linear substructures, whereas $[\Lambda(E^{CE}_{k,i})]^{CE}$ is the matrix of the squared angular frequencies for the NNMs of the nonlinear connecting element at a given energy level $E^{CE}_{k,i}$, computed using the HB method. A unique set of generalized modal coordinates $\{z\}$ can be defined using the following transformation of coordinates:

$$\{y\} = [\tilde{L}]_{k,i}\{z\} \tag{3.7}$$

where $[\tilde{L}]_{k,i}$ depends on the energy level. Thus, substituting Eq. (3.7) in Eq. (3.5) and pre-multiplying by $[\tilde{L}]_{k,i}^T$ becomes

$$[\tilde{M}]_{k,i}\{\ddot{z}\} + [\tilde{K}]_{k,i}\{z\} = \{0\} \tag{3.8}$$

where the matrices $[\tilde{M}]_{k,i}$ and $[\tilde{K}]_{k,i}$ are updated at each iteration. The system in Eq. (3.8) represents a linear approximation of the nonlinear coupled system at e_k. Thus, it is possible to compute the eigenvalues–eigenvectors pairs representing the resonance angular frequency ω_r and the mode shape $\{\xi\}_r$ expressed in generalized modal coordinates $\{z\}$ of the r-th mode. This result can be brought back in the physical domain by performing an inverse transformation according to Eqs. (3.3) and (3.7) to obtain the corresponding mode shape $\{\chi\}_{k,i}$.

However, the predicted mode shape is generally associated with an energy different from e_k. Thus, the predicted vector $\{\chi\}_{k,i}$ is scaled by a coefficient α:

$$\{\hat{\chi}\}_{k,i} = \alpha\{\chi\}_{k,i} \tag{3.9}$$

so that the energy level associated with $\{\hat{\chi}\}_{k,i}$ is forced to be equal to e_k, i.e.

$$-\omega_{k,i}^2\alpha^2\frac{\{\chi\}_{k,i}^T[M]\{\chi\}_{k,i}}{2} + \alpha^2\frac{\{\chi\}_{k,i}^T[K]\{\chi\}_{k,i}}{2} + \int_0^{\alpha\{\chi\}_{k,i}} f_{nl}(\alpha\{\chi\}_{k,i})\, d(\alpha\{\chi\}_{k,i}) = e_k \tag{3.10}$$

and the energy distribution vector $\{\hat{E}\}_{k,i}$ at iteration i can be computed. The iterative algorithm stops when the difference between the energy distribution at two consecutive iterations is below a certain tolerance, i.e.

$$\frac{\left|\{\hat{E}\}_{k,i} - \{\hat{E}\}_{k,i-1}\right|_2}{\left|\{\hat{E}\}_{k,i-1}\right|_2} < \varepsilon \tag{3.11}$$

If Eq. (3.11) is not satisfied, the initial energy for Eqs. (3.4)–(3.8) is updated as

$$\begin{Bmatrix} E_{k,i+1}^A \\ E_{k,i+1}^{CE} \\ E_{k,i+1}^B \end{Bmatrix} = \lambda \begin{Bmatrix} \hat{E}_{k,i}^A \\ \hat{E}_{k,i}^{CE} \\ \hat{E}_{k,i}^B \end{Bmatrix} + (1-\lambda) \begin{Bmatrix} E_{k,i}^A \\ E_{k,i}^{CE} \\ E_{k,i}^B \end{Bmatrix} \tag{3.12}$$

where λ is a weighting factor ranging from 0 to 0.5 to improve the numerical stability of the iterative procedure.

Eventually, the vector $\{E\}_{k,i+1}$ is normalized to enforce that the total energy is always equal to e_k, i.e.

$$\left\{\tilde{E}\right\}_{k,i+1} = \{E\}_{k,i+1}\sqrt{\frac{e_k}{E_{k,i+1}^A + E_{k,i+1}^{CE} + E_{k,i+1}^B}} \tag{3.13}$$

3.2.2 Estimation of the Initial Guess

The iterative algorithm described in Sect. 3.2.1 requires the initial guess for the energy distribution at which the matrix $[R]$ is computed for $i = 1$. The initial guess is chosen to be close to the real solution for every energy level e_k, and this is achieved in two different ways, whether $k \leq 2$ or $k > 2$. In the first case, the corresponding energy levels e_1 and e_2 are chosen to be very low, such that LNMs and NNMs for the coupled system basically coincide in frequency and mode shape. The energy distribution is then estimated by computing the amplitude of the mode shape of the coupled linear system such that its energy is equal to e_k and computing how it is distributed among each linear substructure. In the second case, instead, the solution obtained at the energy levels e_{k-2} and e_{k-1} is used to predict both the mode shape and the energy distribution at e_k. In fact, if the energy increment among three consecutive energy levels is low, it is reasonable to assume that the corresponding solutions are similar.

3.3 Results

The NLCP is used to couple two beams connected through two nonlinear springs, as shown in Fig. 3.3. Each beam is modeled using 20 Euler–Bernoulli finite elements of equal length. The upper beam is unconstrained, while the lower beam is clamped at both ends. The axial degrees of freedom are neglected, thus, in a 2D domain, each node has two degrees of freedom: the vertical displacement and the rotation.

The beams are connected together through a softening spring, described by Eq. (3.14), and an hardening spring, described by Eq. (3.15)

$$k_s = k_\infty + k_{nl,s}\frac{\text{atan}(\beta x)}{x} \tag{3.14}$$

$$k_h = k_{l,h} + k_{nl,h}x^2 \tag{3.15}$$

where $k_\infty = 1000$ N/m and $k_{l,h} = 1000$ N/m are the coefficients of the constant terms, $k_{nl,s} = 1$ N/m and $k_{n,h} = 5\cdot10^6$ N/m^3 are the coefficients of the nonlinear terms, and $\beta = 2000$ m^{-1} is a scaling coefficient of the softening spring. The mechanical properties and dimensions of the beams are summarized in Table 3.1. Note that the two beams have different thickness in order to avoid equal displacement fields at a given frequency.

The first six NNMs of the coupled structure are computed using the NLCP, and the results can be compared to those obtained by applying the HB method with one harmonic using a Frequency–Energy Plot (FEP), as shown in Fig. 3.4a. The curves, describing the variation of the resonance frequency as function of the energy obtained using the NLCP, manage to retrace the results provided by the HB method. Note that the presence of a softening and a hardening spring leads to a different trend of variation of the resonance frequency of each NNM. In fact, it is evident from Fig. 3.4a that the resonance frequency of the second mode decreases as the energy increases, while the opposite occurs for the third and fourth mode. Regarding the fifth mode, instead, the resonance frequency decreases up to ≈ 4 J and then it starts increasing. This non-trivial behavior is successfully retraced by the NLCP, with a maximum deviation in frequency < 0.6 %, as shown in Fig. 3.4b. Figure 3.5 shows, for each mode, how the energy is distributed among the substructures and the stiffness variation of each connecting element at each energy level. It is evident that, as the energy increases, the stiffness of the left connecting element tends to k_∞, while the one of the right element increases according to Eq. (3.15). In the first mode, the upper beam is the substructure that stores almost all the energy available. For the other modes, the change in the energy distribution occurs always in correspondence with the variation of the nonlinearity of the softening connecting element. Furthermore, the increase of the

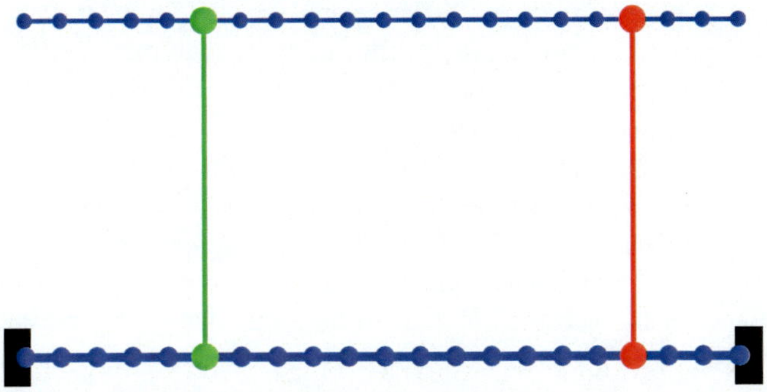

Fig. 3.3 Structure of the system. The horizontal lines represent the beam, the vertical green line represents the softening spring whereas the vertical red line represents the cubic spring, and the two black blocks at the lower beam ends represent the clamps

Table 3.1 Values of the physical and geometrical parameters of the two upper and lower beam

L_{up}	w_{up}	t_{up}	L_{low}	w_{low}	t_{low}		ρ	E	n_{up}	n_{low}
1000	10	2	1000	10	4	mm	7800 kg/m^3	205e9 Pa	21	21

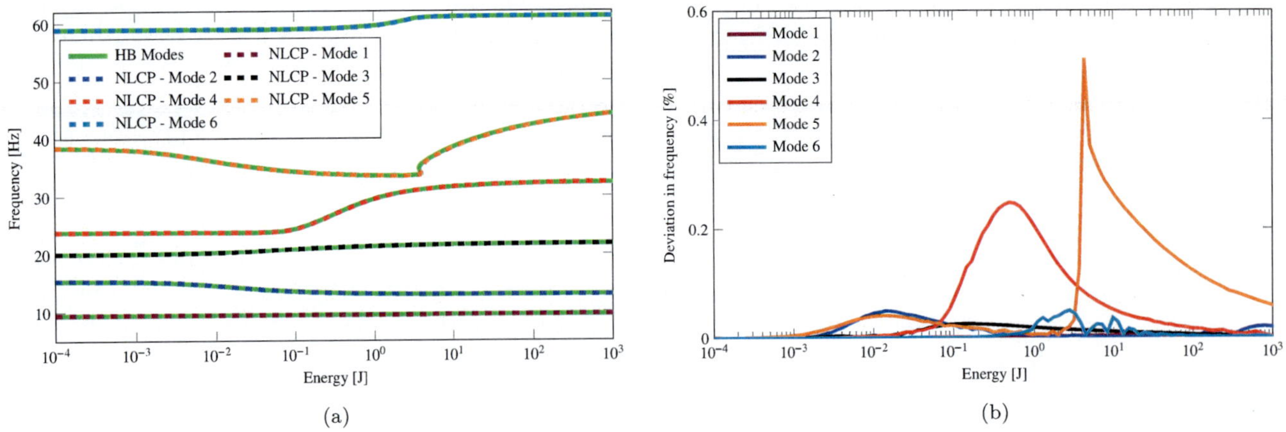

Fig. 3.4 Comparison between the Results obtained using the NLCP and the HB method with one harmonic. (**a**) FEP of the first six NNMs of the coupled structure. (**b**) Deviation in frequency between the results of NLCP and the HB method

resonance frequency of the fifth mode happens at the same energy level where there is an abrupt change in the energy content of the hardening connecting element.

The results shown in Figs. 3.4 and 3.5 are obtained using full modal bases for the modal reduction of each beam, i.e. 42 modes for the upper one and 38 modes for the lower one. However, it is interesting to evaluate the effects of modal truncation either on the resonance frequencies and on the mode shapes of the system. Thus, the NLCP is performed considering a truncated modal basis for each beam, corresponding to five truncation frequencies, as summarized in Table 3.2. The curves describing how the deviation in frequency varies as function of the energy are shown in Fig. 3.6. The curves show that, as the truncation frequency increases, the deviation in frequency decreases. Furthermore, there is also a dependency on the energy at which the system is evaluated, particularly evident for the third, fourth, and fifth mode. Regarding the fifth mode shown in Fig. 3.6e, it is possible to see that the deviation in frequency first decreases and then starts increasing in correspondence with the same energy level at which there is the increase in the resonance frequency.

The correlation between the mode shapes, instead, can be evaluated by computing the Modal Assurance Criterion (MAC) [24] for the five modes at a given energy level, using the HB results as reference:

$$\mathrm{MAC}(\{\chi\}_{HB}, \{\chi\}_{CP}) = \frac{|\{\chi\}_{HB}^T \{\chi\}_{CP}|^2}{(\{\chi\}_{HB}^T \{\chi\}_{HB})(\{\chi\}_{CP}^T \{\chi\}_{CP})} \tag{3.16}$$

For each mode, the MAC is evaluated at two different energy levels, summarized in Table 3.3. Figure 3.7 shows how the MAC varies for the five modes at each truncation frequency at the chosen energy levels. Note that the blue bars in the graphs are referred to the MAC of the complete system, and highlight how the increase in the truncation frequency leads to an overall increase in the MAC number. In the present case, 250 Hz seems to be the lowest truncation frequency that guarantees that the mode shapes computed using the NLCP are not much affected by modal truncation. Furthermore, the partial Modal Assurance Criterion (pMAC) [25] is computed for the upper and lower beams and for the nonlinear connections to evaluate the partial correlation of the mode shapes (orange, yellow, and purple bars in the histograms of Fig. 3.7, respectively). If the pMAC referred to one substructure is low, it indicates the portion of the modal vector which gives poor correlation between the mode shapes. It is interesting to note that the MAC can be lower than the pMACs at the same energy level and truncation frequency: it happens, for example, for the 2nd mode at high energy at 25 Hz, for the 3rd mode at low energy at 25 Hz and for the 5th mode at low energy at 50 Hz. This indicates that the poor correlation highlighted by the MAC depends either on the partial correlation between the substructures or on how the energy is distributed among the substructures. Considering for example the 5th mode, it is possible to compare the mode shapes and the energy distribution among the substructures at each truncation frequency at low and high energy, as shown in Fig. 3.8. Note that the mode shapes computed at 50 Hz are not correct, and this is also highlighted from the energy distribution in Fig. 3.8b–d. In fact, when the truncation frequency is 50 Hz, the energy associated with the hardening spring is much higher than the value it should be, while the other substructures are associated with lower energy. Conversely, when the truncation frequency increases, the energy distribution tends to the reference value and the mode shape basically coincides with the one provided by the HB method. It is also interesting to highlight that the mode shapes at low and high energy are different from each other, as it is evident from Fig. 3.8a–c.

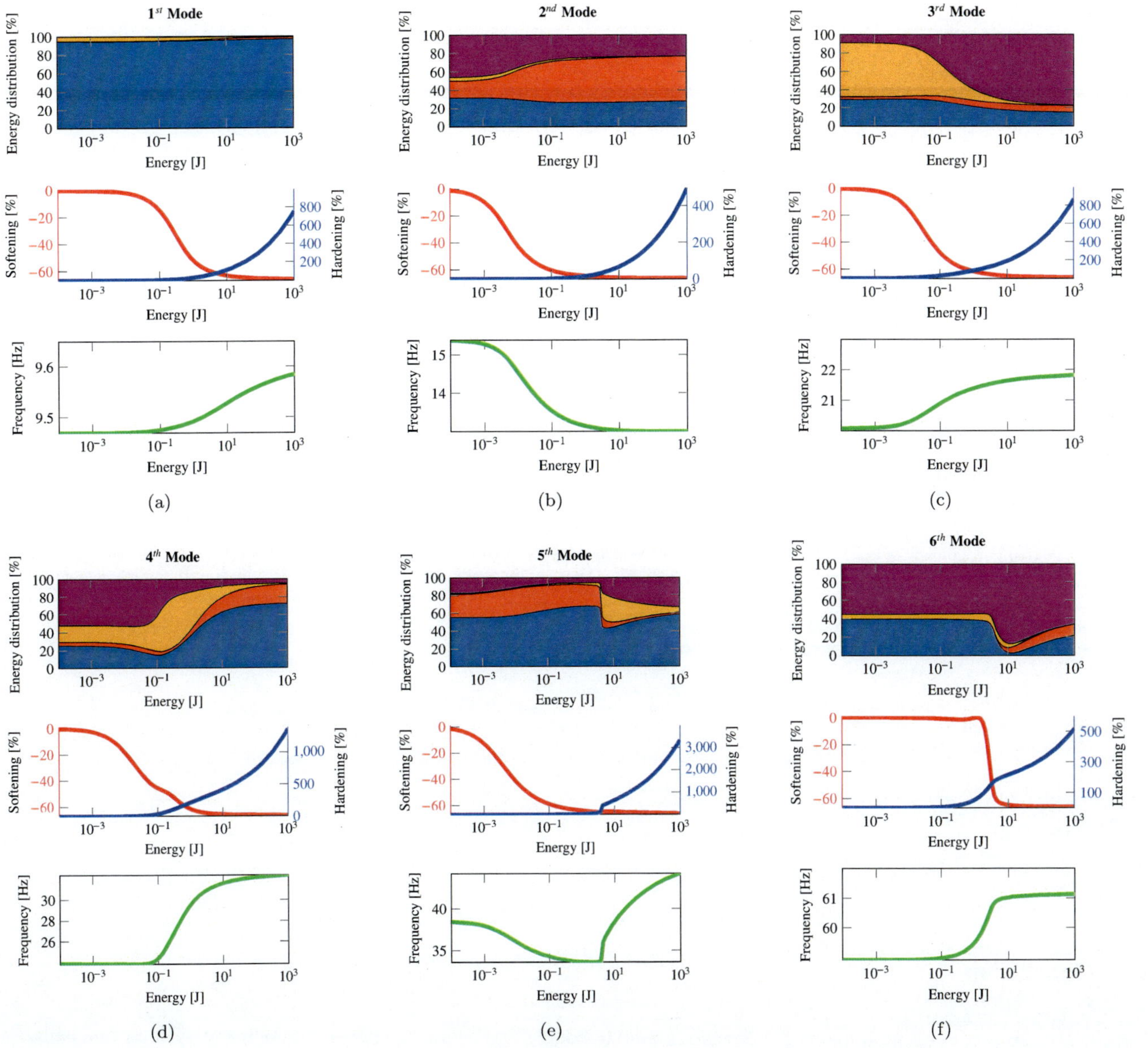

Fig. 3.5 Energy distribution: upper beam, left softening connecting element, right hardening connecting element, lower beam. (**a**) Energy distribution among the substructures and nonlinear contribution of each spring for the 1st mode. (**b**) Energy distribution among the substructures and nonlinear contribution of each spring for the 2nd mode. (**c**) Energy distribution among the substructures and nonlinear contribution of each spring for the 3rd mode. (**d**) Energy distribution among the substructures and nonlinear contribution of each spring for the 4th mode. (**e**) Energy distribution among the substructures and nonlinear contribution of each spring for the 5th mode. (**f**) Energy distribution among the substructures and nonlinear contribution of each spring for the 6th mode

Table 3.2 Number of modes included for each modal truncation case

Truncation frequency [Hz]	25	50	100	250	500	No truncation
N° of modes used	4	5	8	12	17	80

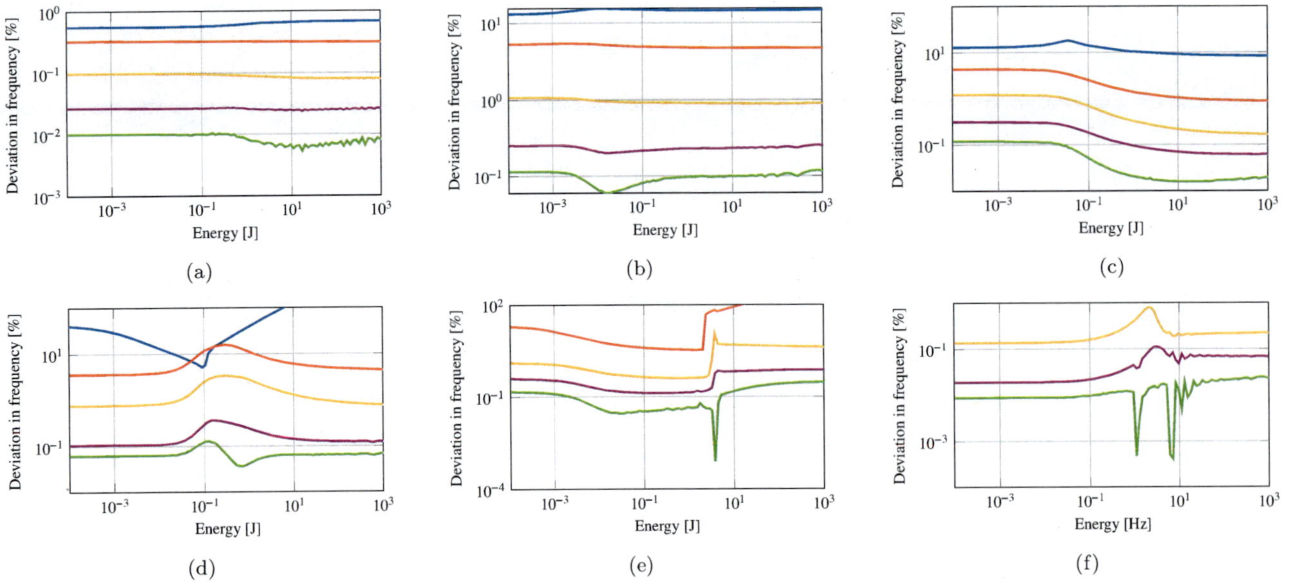

Fig. 3.6 Deviation in frequency of the 6 modes for each truncation frequency: – 25 Hz, – 50 Hz, – 100 Hz, – 250 Hz, – 500 Hz. (**a**) Deviation in frequency for the 1st Mode with linear frequency equal to 9.5 Hz. (**b**) Deviation in frequency for the 2nd Mode with linear frequency equal to 15.4 Hz. (**c**) Deviation in frequency for the 3rd Mode with linear frequency equal to 20.1 Hz. (**d**) Deviation in frequency for the 4th Mode with linear frequency equal to 23.9 Hz. (**e**) Deviation in frequency for the 5th Mode with linear frequency equal to 38.5 Hz. (**f**) Deviation in frequency for the 6th Mode with linear frequency equal to 58.9 Hz

Table 3.3 Energy levels for each mode

	1st Mode	2nd Mode	3rd Mode	4th Mode	5th Mode	6th Mode
Low energy [J]	0.0015	0.0015	0.0015	0.02	2.3	0.002
High energy [J]	1610	974	285	2100	2500	84

3.4 Conclusion

The substructuring procedure in the modal domain is used to couple two beams connected through two springs, one being softening and the other being hardening. The method manages to retrace the variation of the resonance frequency as function of the total energy stored in the system. It is also able to capture the effects that the deformation of the springs have on the resonance frequency, as it happens with the fifth mode where the resonance frequency decreases and then increases as the energy grows. The effects due to modal truncation have been evaluated by comparing the resonance frequencies and the mode shapes to the results obtained using the HB method with one harmonic. As expected, the deviation in the resonance frequency increases as the truncation frequency decreases. However, it varies also depending on the energy level at which the system is evaluated, especially if the variation of the resonance frequency is large. The modal truncation affects also the mode shapes depending on the energy level at which the system is evaluated, as highlighted from the MAC. In particular, it is possible to compute how the energy is distributed among the substructures to verify if the chosen truncation frequency is suitable to describe the dynamics of the model at the given energy level. The results show that the substructuring procedure is able to determine the dynamics of linear systems connected through nonlinear connections, having either a hardening or softening behavior.

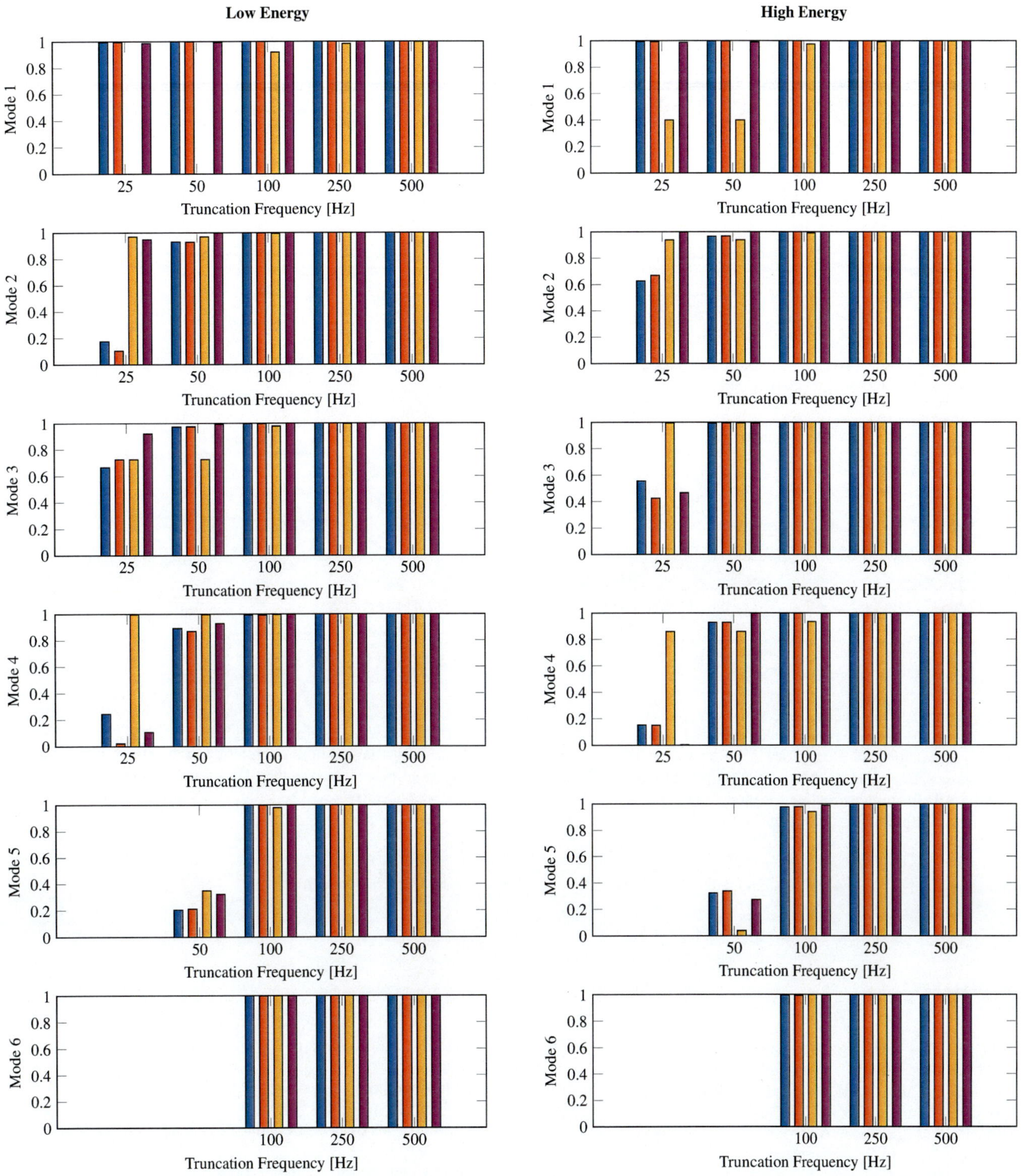

Fig. 3.7 MAC number for the 5 modes for each truncation frequency at low and high energy: ■ MAC, ■ pMAC for the upper beam, ■ pMAC for the lower beam, ■ pMAC for the nonlinear connections

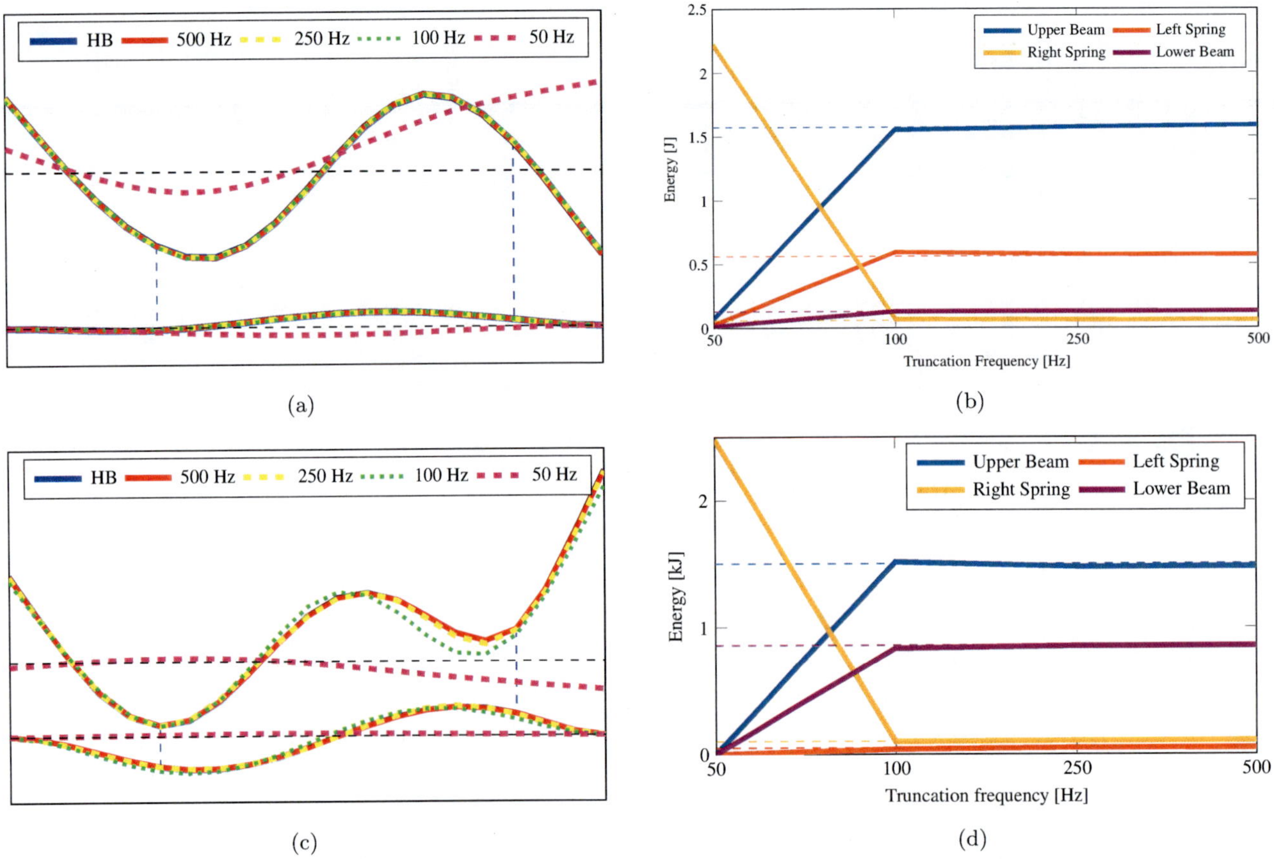

Fig. 3.8 Variation of the mode shape and energy distribution among substructures for mode 5 at low and high energy at different truncation frequencies. (**a**) Mode shapes for mode 5 at 2.3 J evaluated at different truncation frequencies. Undeformed model - -, springs' location - -. (**b**) Energy distribution for mode 5 at 2.3 J evaluated at different truncation frequencies. The dashed lines are referred to the HB results for the corresponding substructure. (**c**) Mode shapes for mode 5 at 2500 J evaluated at different truncation frequencies. Undeformed model - -, springs' location - -. (**d**) Energy distribution for mode 5 at 2500 J evaluated at different truncation frequencies. The dashed lines are referred to the HB results for the corresponding substructure

Acknowledgments This research is supported by University of Rome La Sapienza and University of L'Aquila.

References

1. de Klerk, D., Rixen, D.J., Voormeeren, S.N.: General framework for dynamic substructuring: history, review, and classification of techniques. AIAA J. **46**(5) (2008)
2. Brunetti, J., Culla, A., D'Ambrogio, W., Fregolent, A.: Experimental Dynamic Substructuring of the Ampair Wind Turbine Test Bed, vol. 1, pp. 15–26. Springer, Cham (2014)
3. Brunetti, J., Culla, A., D'Ambrogio, W., Fregolent, A.: Selection of Interface DoFs in Hub-Blade(s) Coupling of Ampair Wind Turbine Test Bed, vol. 2, pp. 167–178. Springer, Cham (2014)
4. Kalaycioğlu, T., Özgüven, H. N.: Harmonic response of large engineering structures with nonlinear modifications. In: Proceedings of the 8th International Conference on Structural Dynamics, EURODYN, pp. 3623–3629 (2011)
5. Kalaycıoğlu, T., Özgüven, H.N.: Nonlinear structural modification and nonlinear coupling. Mech. Syst. Signal Process. **46**(2), 289–306 (2014)
6. Kalaycıoğlu, T., Özgüven, H.N.: FRF decoupling of nonlinear systems. Mech. Syst. Signal Process. **102**, 230–244 (2018)
7. Özgüven, H.N.: Structural modifications using frequency response functions. Mech. Syst. Signal Process. **4**(1), 53–63 (1990)
8. Vander Velde, W.E.: Multiple-Input Describing Functions and Nonlinear System Design. McGraw-Hill, New York (1968)
9. Kuether, R.J., Allen, M.S., Hollkamp, J.J.: Modal substructuring of geometrically nonlinear finite element models with interface reduction. AIAA J., 1695–1706 (2017)
10. Mahdiabadi, M.K., Bartl, A., Xu, D., Tiso, P., Rixen, D.J.: An augmented free-interface-based modal substructuring for nonlinear structural dynamics including interface reduction. J. Sound Vib. **462**, 114915 (2019)
11. Hollkamp, J.J., Gordon, R.W.: Reduced-order models for nonlinear response prediction: implicit condensation and expansion. J. Sound Vib. **318**(4–5), 1139–1153 (2008)

12. Kuether, R.J., Allen, M.S.: Nonlinear modal substructuring of systems with geometric nonlinearities. In: 54th AIAA/ASME/ASCE/AHS/ASC Structures, Structural Dynamics, and Materials Conference, p. 1521 (2013)
13. Rosenberg, R.M.: Normal modes of Nonlinear dual-mode systems. J. Appl. Mech. **27**(2), 263–268 (1960)
14. Kerschen, G., Peeters, M., Golinval, J., Vakakis, A.F.: Nonlinear normal modes, Part I: a useful framework for the structural dynamicist. Mech. Syst. Signal Process. **23**(1), 170–194 (2009)
15. Lacayo, R., Pesaresi, L., Groß, J., Fochler, D., Armand, J., Salles, L., Schwingshackl, C., Allen, M., Brake, M.: Nonlinear modeling of structures with bolted joints: a comparison of two approaches based on a time-domain and frequency-domain solver. Mech. Syst. Signal Process. **114**, 413–438 (2019)
16. Balaji, P.S., Rahman, M.E., Moussa, L., Lau, H.H., Wire rope isolators for vibration isolation of equipment and structures–a review. In: IOP Conference Series: Materials Science and Engineering, vol. 78, p. 012001. IOP Publishing, Bristol (2015)
17. Saeed, Z., Firrone, C.M., Berruti, T.M.: Joint identification through hybrid models improved by correlations, Journal of Sound and Vibration, **494** (2021)
18. Krack, M., Panning-von Scheidt, L., Wallaschek, J.: On the computation of the slow dynamics of nonlinear modes of mechanical systems. Mech. Syst. Signal Process. **42**(1–2), 71–87 (2014)
19. Latini, F., Brunetti, J., D'Ambrogio, W., Fregolent, A.: Substructures' coupling with nonlinear connecting elements. Nonlinear Dyn. **99**(2), 1643–1658 (2020)
20. Brunetti, J., D'Ambrogio, W., Fregolent, A., Latini, F.: Substructuring using NNMS of nonlinear connecting elements. In: Lecture Notes in Mechanical Engineering, pp. 1426–1440 (2020)
21. Latini, F., Brunetti, J., Kwarta, M., Allen, M.S., D'Ambrogio, W., Fregolent, A.: Experimental results of nonlinear structure coupled through nonlinear connecting elements. In: Proceedings of ISMA 2020 – International Conference on Noise and Vibration Engineering and USD 2020 – International Conference on Uncertainty in Structural Dynamics (2020)
22. Falco, M., Mahdiabadi, M.K., Rixen, D.J.: Nonlinear substructuring using fixed interface nonlinear normal modes. In: Dynamics of Coupled Structures, vol. 4, pp. 205–213. Springer, Berlin (2017)
23. Latini, F., Brunetti, J., D'Ambrogio, W., Allen, M.S., Fregolent, A.: Nonlinear substructuring in the modal domain: numerical validation and experimental verification in presence of localized nonlinearities, Nonlinear Dynamics, **140**, 1043–1067 (2021)
24. Allemang, R.J., Brown, D.L.: A correlation coefficient for modal vector analysis. In: Proceedings of the 1st International Modal Analysis Conference, vol. 1, pp. 110–116. SEM, Orlando (1982)
25. Heylen, W., Janter, T.: Extensions of the modal assurance criterion. J. Vib. Acoust. **112**(4), 468–472 (1990)

Chapter 4
On the Coupling of Reduced Order Modeling with Substructuring of Structural Systems with Component Nonlinearities

Konstantinos Vlachas, Konstantinos Tatsis, Konstantinos Agathos, Adam R. Brink, Dane Quinn, and Eleni Chatzi

Abstract The emergence of digital virtualization has brought Reduced Order Models (ROM) into the spotlight. A successful reduced order representation should allow for modeling of complex effects, such as nonlinearities, and ensure validity over a domain of inputs. Parametric Reduced Order Models (pROMs) for nonlinear systems attempt to accommodate both previous requirements (Benner et al., SIAM Rev 57(4), 1–14, 2015). Our work addresses a physics-based reduced representation of structural systems with localized nonlinear features. Via implementation of the approach described in Quinn (J Sound Vib 331(1), 81–93, 2012), we achieve a substructuring formulation similar to Component Mode Synthesis. However, instead of individual component modes, reduction modes of a global nature are obtained from the corresponding linear monolithic system. In turn, this technique allows for a divide and conquer strategy that naturally couples the response between linear and nonlinear subdomains. A pROM able to exploit this modular formulation is developed as a next step. The framework treats each component independently, and individual projection bases are assembled by applying a Proper Orthogonal Decomposition (POD) on snapshots of each subdomain's response. Parametric dependencies on the boundary conditions and the structural and excitation traits are injected into the pROM utilizing clustering, following the methodology in Vlachas et al. (Vlachas, K., Tatsis, K., Agathos, K., Brink, A.R., Chatzi, E.: A local basis approximation approach for nonlinear parametric model order reduction. Journal of Sound and Vibration 502, 116055, 2021). To this end, a modified strategy is employed, relying on the Modal Assurance Criterion as a measure to indicate optimal training samples while defining regions with similar underlying dynamics on the domain of parametric inputs. A numerical case study of a 3D wind turbine tower featuring material nonlinearities exemplifies our approach. The derived pROM offers an accelerated approximation of the underlying high fidelity response and can be utilized for numerous tasks, including vibration control, residual life estimation, and condition assessment of hotspot locations.

Keywords Parametric ROM · Nonlinear · Substructuring · Adaptive sampling

4.1 Introduction

Dealing with real-life, true scale dynamical systems often requires incorporation of high complexity models, comprising multiple components. Dynamic substructuring addresses this issue and encompasses several methodologies that aim to decompose the system into individual entities, treat them separately, and finally assemble them properly to obtain the global response. A detailed overview of the notion is provided in [4], whereas recent advances are discussed in [5, 6], including

K. Vlachas (✉) · K. Tatsis · E. Chatzi
Department of Civil, Environmental and Geomatic Engineering, ETH Zurich, Zurich, Switzerland
e-mail: vlachas@ibk.baug.ethz.ch; tatsis@ibk.baug.ethz.ch; chatzi@ibk.baug.ethz.ch

K. Agathos
College of Engineering, Mathematics and Physical Sciences, University of Exeter, Exeter, UK
e-mail: K.Agathos@exeter.ac.uk

A. R. Brink
Component Sciences and Mechanics, Sandia National Laboratories, Albuquerque, NM, USA
e-mail: arbrink@sandia.gov

D. Quinn
Department of Mechanical Engineering, The University of Akron, Akron, OH, USA
e-mail: quinn@uakron.edu

M. S. Allen et al. (eds.), *Dynamic Substructures, Volume 4*, Conference Proceedings of the Society for Experimental Mechanics Series, https://doi.org/10.1007/978-3-030-75910-0_4

comparative assessments, experimental techniques, and industrial applications. Most of the methodologies that adhere to dynamic substructuring can approximate each component's underlying physics separately by employing some form of fundamental modes. These correlate to the established notion of eigenmodes, in that they usually represent characteristic (deformed) configurations of the component (or the structure) and compress the response information in a few vectors able to explain and efficiently reproduce the high fidelity complex behavior [7]. In a linear setting, this assumption may hold as long as these modes are appropriately formulated. However, local phenomena may dominate the response when nonlinearities are present, thus violently changing dynamics and rendering this simplification ineffective.

Therefore, substructural representations of nonlinear systems require additional treatment compared against traditional methodologies [6]. For example, Wu et al. [8] augmented Rubin substructuring with information captured by modal derivatives to model a three-dimensional wind turbine blade with geometrical nonlinearities. On a similar setup, Schoneman et al. [9] coupled Component Mode Synthesis (CMS) with the constraint modes approach to address the challenges associated with geometrically nonlinear panel and frame assemblies. In this context, though, nonlinear normal modes (NNMs) seem to be the most referenced technique for handling nonlinearities in a substructural formulation [10]. The respective computational framework attempts to adapt the notion of modes explained in the previous paragraphs in a nonlinear setting, and it has been proven effective in a variety of applications, including resonance prediction [11] or models with nonlinear friction force effects [12].

The notion of exploiting dynamic substructuring to formulate a reduced order representation is not new. Dynamic substructuring relies on modes that capture the overall dynamics in a set of reduced coordinates, thus implying a form of dimensionality reduction [6]. For instance, Gruber et al. [13] derived a ROM for non-classically damped linear systems utilizing dual CB-CMS, whereas linear parametric problems are addressed in [14], using CB-CMS modes interpolation on the proper subspace.

In a nonlinear context, Weeger et al. [15] suggested a ROM based on modal derivatives, able to substitute all nonlinear terms via polynomials of the reduced state variables, whereas Joannin et al. [16] coupled NNMs with modal synthesis to address nonlinear friction phenomena. As an alternative methodology to NNMs, modes derived through Proper Orthogonal Decomposition (POD) have been coupled with CMS in a substructural formulation in [17, 18]. Through POD, a ROM can harvest information from a system's response and approximate the solution manifold spanning the dynamic behavior of a component [1]. This enables an accurate and efficient projection-based reduction.

In this work, we derive a physics-based ROM able to treat parametric dependencies in the global system while the reduced treatment of each component takes place separately. To achieve this, we couple the POD-based algorithm implemented in [3] with the substructural formulation in [2, 19]. The representation in [2] is critical for the parametric ROM. By decomposing the system at any time into an ideal and a deviatoric representation, the ROM is fed with response information on both a global model and a localized component level. On a global scale, the POD modes of the ideal system capture the monolithic assembly dynamics without the presence of localized features. These features are in turn approximated in a component-wise manner by assembling suitable POD deviatoric subspaces, thus not requiring any additional treatment such as mode expansion [18].

Further, our approach addresses the necessity of efficiently exploring the parametric domain of a ROM [20] during the sampling of the high fidelity training states. This is achieved by utilizing greedy sampling techniques with ROM's performance error indicators [21]. In this work, the Modal Assurance Criterion [22] is employed, relating and indirectly comparing the dynamics captured by the POD projection bases of neighboring samples, thus indicating if sampling refinement is required. The derived pROM is evaluated on a 3D case study featuring material nonlinearities on a constitutive level, thus contributing to the available literature of geometrically nonlinear problems [8], nonlinear damping [23], and interfaces or contact modeling [16].

4.2 Methodology

4.2.1 Substructuring of Systems with Component Nonlinearities

Following the methodology in [19], a dynamical system is assumed, comprising two linear components as in Fig. 4.1. These are termed as "exterior regions" and are identified as C_1 and C_2, respectively. The system further comprises an internal component C_s, termed as "isolated region." The exterior regions are assumed to behave linearly, while nonlinearities are present only on the isolated region, represented by $\mathbf{G} \in \mathbb{R}^{n_s}$. Parametric dependencies are represented by vector $\mathbf{p} = [p_1, \ldots, p_\mathrm{m}]^\mathrm{T} \in \Omega \subset \mathbb{R}^m$. The resulting governing equations of the dynamic response read:

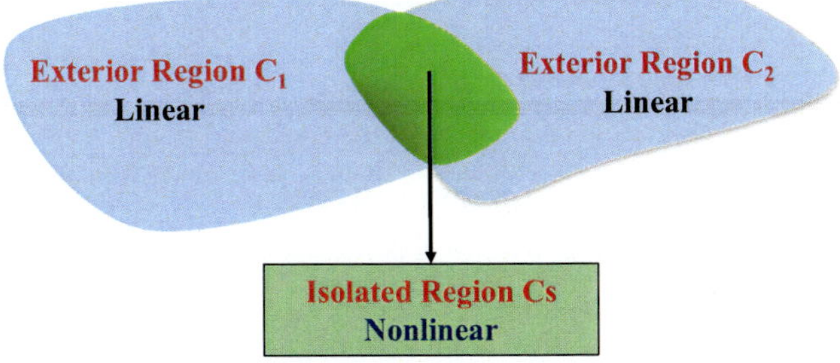

Fig. 4.1 Structural system with localized nonlinear feature, adjusted from [2]

$$\mathbf{M}(\mathbf{p})\ddot{\mathbf{u}}(t) + \mathbf{C}(\mathbf{p})\dot{\mathbf{u}}(t) + \mathbf{K}(\mathbf{p})\mathbf{u}(t) + \mathbf{G}_*(\mathbf{p}) = \mathbf{f}(t, \mathbf{p}), \tag{4.1}$$

$$\mathbf{u} = [\mathbf{u}_1 \quad \mathbf{u}_s \quad \mathbf{u}_2]^{\mathrm{T}}, \quad \mathbf{G}_* = [0 \quad \mathbf{G} \quad 0]^{\mathrm{T}}, \quad n = n_1 + n_s + n_2 \tag{4.2}$$

where $\mathbf{u}(t) \in \mathbb{R}^n$ represents the displacement, $\mathbf{M}, \mathbf{C}, \mathbf{K} \in \mathbb{R}^{n \times n}$ denote the mass, damping, and stiffness matrices, and $\mathbf{f}(t, \mathbf{p}) \in \mathbb{R}^n$ the externally applied excitation. The order of the high fidelity system is n.

The response of the isolated region $\mathbf{u}_s$ is dominated by the presence of nonlinearities, and we assume it can be decomposed as $\mathbf{u}_s = \mathbf{x}+\mathbf{z}$, where $\mathbf{x}$ represents an ideal system to be defined and $\mathbf{z}$ a deviatoric component, capturing the difference between the ideal and the actual representation. This assumption implies that $\mathbf{z}$ is driven by the nonlinear terms $\mathbf{G}$. We express the overall response as $u_x = [\mathbf{u}_1, \mathbf{x}_\alpha, \mathbf{x}_\gamma, \mathbf{x}_\beta, \mathbf{u}_2]^{\mathrm{T}}$, where the subscripts α, β denote the boundaries between components C_1, C_2 and C_s, respectively. By dropping time and parametric dependencies, the motion equations are reformulated as

$$\begin{bmatrix} \mathbf{M}_{11} & \mathbf{M}_{1\alpha} & 0 & 0 & 0 \\ \mathbf{M}_{1\alpha}^{\mathrm{T}} & \mathbf{M}_{\alpha\alpha} & \mathbf{M}_{\gamma\alpha} & 0 & 0 \\ 0 & \mathbf{M}_{\gamma\alpha}^{\mathrm{T}} & \mathbf{M}_{\gamma\gamma} & \mathbf{M}_{\gamma\beta} & 0 \\ 0 & 0 & \mathbf{M}_{\gamma\beta}^{\mathrm{T}} & \mathbf{M}_{\beta\beta} & \mathbf{M}_{2\beta} \\ 0 & 0 & 0 & \mathbf{M}_{2\beta}^{\mathrm{T}} & \mathbf{M}_{22} \end{bmatrix} \begin{bmatrix} \ddot{\mathbf{u}}_1 \\ \ddot{\mathbf{x}}_\alpha \\ \ddot{\mathbf{x}}_\gamma \\ \ddot{\mathbf{x}}_\beta \\ \ddot{\mathbf{u}}_2 \end{bmatrix} + \tilde{\mathbf{C}}\dot{u}_x + \tilde{\mathbf{K}}u_x = \mathbf{f} + \mathbf{R} = \begin{bmatrix} \mathbf{f}_1 - (\mathbf{M}\ddot{\mathbf{z}}_\alpha + \mathbf{C}\dot{\mathbf{z}}_\alpha + \mathbf{K}\mathbf{z}_\alpha) \\ \mathbf{f}_\alpha \\ \mathbf{f}_\gamma \\ \mathbf{f}_\beta \\ \mathbf{f}_2 - (\mathbf{M}\ddot{\mathbf{z}}_\beta + \mathbf{C}\dot{\mathbf{z}}_\beta + \mathbf{K}\mathbf{z}_\beta) \end{bmatrix} \tag{4.3}$$

where the matrices $\tilde{\mathbf{C}}$ and $\tilde{\mathbf{K}}$ are expressed similar to $\tilde{\mathbf{M}}$. With this formulation in place, $\mathbf{R}$ represents the deviatoric force terms driven by the internal response and acting on the exterior regions. Thus, $\mathbf{x}$ captures the response under the additional excitation terms arising from $\mathbf{R}$. To determine the deviatoric terms, the response of the interior region C_s is obtained by

$$\begin{bmatrix} \mathbf{M}_{\alpha\alpha} & \mathbf{M}_{\gamma\alpha} & 0 \\ \mathbf{M}_{\gamma\alpha}^{\mathrm{T}} & \mathbf{M}_{\beta\beta} & \mathbf{M}_{\gamma\beta} \\ 0 & \mathbf{M}_{\gamma\beta}^{\mathrm{T}} & \mathbf{M}_{\beta\beta} \end{bmatrix} \begin{bmatrix} \ddot{\mathbf{z}}_\alpha \\ \ddot{\mathbf{z}}_\gamma \\ \ddot{\mathbf{z}}_\beta \end{bmatrix} + \tilde{\mathbf{C}}_s\dot{z} + \tilde{\mathbf{K}}_s z = \begin{bmatrix} \mathbf{G}_\alpha(\mathbf{x}_\alpha + \mathbf{z}_\alpha, \mathbf{x}_\gamma + \mathbf{z}_\gamma) \\ \mathbf{G}_\gamma(\mathbf{x}_\alpha + \mathbf{z}_\alpha, \mathbf{x}_\gamma + \mathbf{z}_\gamma, \mathbf{x}_\beta + \mathbf{z}_\beta) \\ \mathbf{G}_\beta(\mathbf{x}_\gamma + \mathbf{z}_\gamma, \mathbf{x}_\beta + \mathbf{z}_\beta) \end{bmatrix} \tag{4.4}$$

where $\tilde{\mathbf{C}}_s$ and $\tilde{\mathbf{K}}_s$ are expressed in a substructural formulation similar to $\tilde{\mathbf{M}}_s$. All in all, the mixed (ideal) system in Eq. (4.3) has substituted the governing equations of motion in (4.1) by utilizing the nonlinear component equations in (4.4) to account for the deviatoric response $\mathbf{z} \in \mathbb{R}^{n_s}$.

4.2.2 POD-Based ROM in a Substructural Formulation

The governing equations in Eq. (4.1) can be transformed into a reduced set of coordinates denoting the ROM equations. We assume here that the dynamic response for any given operating parametric sample $\mathbf{p}_{\mathrm{j}}$ spans a low dimensional subspace $S_j \subset \mathbb{R}^n$. As such, the full-order response in Eq. (4.1) can be expressed in the reduced form as

$$\mathbf{u}(t) = \mathbf{V}(\mathbf{p}_{\mathrm{j}})\mathbf{u}_{\mathrm{r}}(t), \tag{4.5}$$

Table 4.1 Algorithmic Process of the pROM. The maximum number of clusters can be used instead of a tolerance during Step 2

Notation:
N_s: Number of training samples, N_t: Number of timesteps
n: Full-order dimension, r: Reduced order dimension
Input:
Parameter vector of training realizations $\mathbf{p} = [\mathbf{p}_1, \mathbf{p}_2, \ldots, \mathbf{p}_{N_s}]$,
Realization $\mathbf{p}_q$ to be approximated
Output: Quantity of interest for realization $\mathbf{p}_q$ estimated with pROM
Training phase
1:**for** k=1,. . . ,N_s **do**
1^a: Simulate high fidelity model for $\mathbf{p} = \mathbf{p}_k$ and obtain displacement time histories $\mathbf{U}_{x,k} \in \mathbb{R}^{n \times N_t}$ (Eq. (4.3)) and $\mathbf{Z}_k \in \mathbb{R}^{n_s \times N_t}$ (Eq. (4.4))
1^b: Perform POD to obtain local bases $\mathbf{V}_{x,k} \in \mathbb{R}^{n \times r_x}$ and $\mathbf{V}_{z,k} \in \mathbb{R}^{n_s \times r_z}$
2: Cluster training samples employing Modal Assurance Criterion (MAC) as a metric.
2^a: Evaluate MAC values between POD bases of training realizations
2^b: Assign training samples to the cluster that its center minimizes the MAC
2^c: Identify maximum MAC on clusters and check if exceeds pre-defined tolerance
2^d: If so, define a new cluster center on the point of maximum MAC
2^e: Refine the sampling domain by adding training states between cluster center(s) and maximum MAC point(s) and repeat from 1 (if needed).
Evaluation phase
1: Identify cluster region of $\mathbf{p}_q$ using nearest neighbors alg. 2: Assume that $\mathbf{V}_{x,q}$ and $\mathbf{V}_{z,q}$ are approximated using the respective bases of the cluster center (or the closest training sample on the cluster)
3: Formulate the reduced order matrices based on Eq. (4.7) for Eqs. (4.3) and (4.4) employing $\mathbf{V}_{x,q}$ and $\mathbf{V}_{z,q}$ respectively
4: Evaluate the pROM approximation of Eq. (4.6) for quantity of interest $\mathbf{Q}_q$

where $\mathbf{u}_r \in \mathbb{R}^r$ and $r << n$ defines the components of the solution in the subspace S. $\mathbf{V}(\mathbf{p}_j)$ denotes the projection basis of the ROM, employed hereafter to obtain the following equivalent, low-order formulation of Eq. (4.1):

$$\mathbf{M}_r(\mathbf{p})\ddot{\mathbf{u}}(t) + \mathbf{C}_r(\mathbf{p})\dot{\mathbf{u}}(t) + \mathbf{K}_r(\mathbf{p})\mathbf{u}(t) + \mathbf{G}_r(*(\mathbf{p}) = \mathbf{f}_r(t, \mathbf{p}), \tag{4.6}$$

$$\text{where} \quad \mathbf{M}_r(\mathbf{p}_j) = \mathbf{V}(\mathbf{p}_j)\mathbf{M}(\mathbf{p}_j)\mathbf{V}(\mathbf{p}_j) \qquad \mathbf{G}_r(\mathbf{p}_j) = \mathbf{V}(\mathbf{p}_j)\mathbf{G}\left(\mathbf{u}(t), \mathbf{p}_j\right) \tag{4.7}$$

The damping and stiffness matrices are projected similar to $\mathbf{M}_r$, whereas the force vector similar to $\mathbf{G}_r$. In deriving an accurate reduced representation, assembling a proper projection basis that can capture the system dynamics is a challenging task. In our work, we apply Proper Orthogonal Decomposition (POD) to collected time histories of the response, hence termed snapshots. By exploiting this technique our parametric ROM is able to approximate a solution manifold at any given input sample state of a family of parametric training realizations.

Our approach assembles two low-order subspaces, allowing us to exploit the substructural formulation described in this section in a ROM context and capture response information on a global and a localized scale. Following Eq. (4.3), a global system projection basis is assembled from $\mathbf{u}_x$, capturing elements of the dynamic behavior of the ideal system. The deviatoric component $\mathbf{z}$ of the response, obtained through Eq. (4.4), is treated separately. This process implies that an individual subspace is assembled for the nonlinear component, addressing local dynamics in an isolated manner. The overall algorithmic procedure implemented in this study is presented in detail in Table 4.1. The greedy sampling strategy is presented next.

4.2.3 Modal Assurance Criterion for a Greedy Sampling Approach

As described in Table 4.1, the Modal Assurance Criterion (MAC) [22], or vector cosine or cosine similarity measure, is utilized as a clustering and greedy sampling metric during the training phase of the parametric Reduced Order Model. MAC is defined as a scalar constant representing a measure of consistency between two modal vectors $\boldsymbol{\phi}_r$ and $\boldsymbol{\phi}_s$:

$$\text{MAC}(\boldsymbol{\phi}_r, \boldsymbol{\phi}_s) = \frac{|\boldsymbol{\phi}_r{}^T \boldsymbol{\phi}_s|^2}{(\boldsymbol{\phi}_r{}^T \boldsymbol{\phi}_r)(\boldsymbol{\phi}_s{}^T \boldsymbol{\phi}_s)} \tag{4.8}$$

It assumes values spanning from zero to one, representing the degree of consistent correspondence. This implies that, if the two vectors exhibit a consistent linear relationship, the respective MAC measure should approach unity. In the system identification context, this consistency provides a form of confidence factor when evaluating modal vectors arising from different sources [22]. Since POD captures the local dynamics of the training realizations and the columns of the derived projection bases represent POD modal vectors, MAC can be utilized as a qualitative comparative measure to assess if two bases approximate the same solution manifold. Thus, relating the respective bases using MAC delivers an indirect assessment of the ability of the cluster center to accurately approximate neighboring dynamics.

4.3 Results

Following the outline of the implemented approach, this section presents the numerical evaluation steps. First, the efficiency of the MAC employed as a clustering metric is demonstrated in an academic example of a 3D, two-story building with nonlinear parametric links under earthquake excitation. This initial example focuses on evaluating the proposed sampling methodology of Table 4.1 without any substructuring methodology before assessing the performance of the pROM in a complex case study. The second example features a 3D real-scale wind turbine tower with a plasticity domain and parametric dependencies pertaining to the excitation and boundary conditions' characteristics. Both cases studies are adjusted from [3, 24].

4.3.1 *Two-Story Frame with Hysteretic Links*

All beam elements of the frame structure feature a rectangular cross section (40 cm × 40 cm), made of reinforced concrete. The Young modulus is $E = 21$ GPa and the density is $\rho = 2400\frac{\text{kg}}{\text{m}^3}$. The length of each frame is $l = 7.5$ m, the width is $w = 5$ m, and the height $h = 3.2$ m. The structure is excited using the Kobe accelerogram, assuming an earthquake ground motion with an angle of $pi/4$. Additionally, 4% mass proportional damping is assumed. All degree of freedom-to-degree of freedom couplings on every node of the structure are modeled based on a nonlinear force-displacement law. The restoring force terms z are computed employing the Bouc Wen hysteretic model. The respective Equation is

$$\dot{\mathbf{z}} = \mathbf{A}\dot{x} - \beta|\dot{x}|\mathbf{z}|\mathbf{z}|^{w-1} - \gamma\dot{x}|\mathbf{z}|^{w}, \quad with \quad \mathbf{z}_{max} = \left(\frac{\mathbf{A}}{\beta+\gamma}\right)^{\frac{1}{w}}, \tag{4.9}$$

where x denotes the displacement and $\mathbf{A}$ controls the amplitude of the hysteresis curve. Parameters β and γ dictate its shape and steepness. Dependencies on the parameters of the hysteretic links, namely the amplitude parameter $A \in [2.2, 6.6]$, and $z_{max} \in [0.22.2.0]$, are injected into the pROM, thus evaluating its ability to approximate different effects denoted by shape and amplitude changes in the hysteresis curves. After performing a parametric study for the projection bases order in Fig. 4.2a, similar to [24], a reduced dimension of $r = 16$ is chosen. Then, we sample our domain using latin hypercube sampling (lhs).

Figure 4.2b presents an initial evaluation of the pROM if only the midpoint sample was used during training. The respective error measures are summarized in Table 4.2. Assuming no prior knowledge of the dynamics across the parametric space, the MAC strategy assembles an initial training set using ten random samples (latin hypercube sampling design). The training phase of Table 4.1 is initiated as next. Figure 4.2c depicts the cluster centers after two iterations and the respective regions if the validation set was to be clustered. Considering the random sampling nature of the process, the two defined centers are close to high error regions, based on Fig. 4.2b. This implies that MAC-driven clustering can locate the regions where additional samples are required properly.

Based on a pre-defined tolerance, the MAC clustering converged after six iterations. Regarding the random nature of the initial sampling, the pROM has been validated for several seeds to ensure that, although the order the centers are defined may change, the converged solution remains similar. The respective accuracy performance of the pROM evaluated on the validation set is depicted in Fig. 4.3a. In addition, Fig. 4.3c depicts a structured grid sampling, after [3, 24]. The sampling

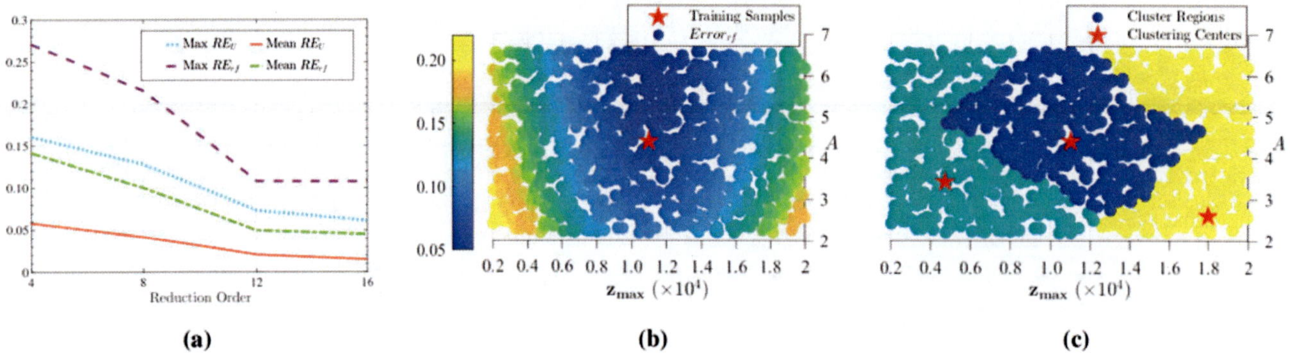

Fig. 4.2 Basis order and example demonstration on effectiveness of MAC clustering. (**a**) Reduced order selection. (**b**) Initial approximation error. (**c**) Clusters after 2 iterations

Table 4.2 Evaluation of the pROM performance. Statistical measures of the error across the domain are presented

	Displacement accuracy		Restoring Forces accuracy	
	Median (σ %)	Maximum	Median (σ %)	Maximum
Initial approximation	1.56% (0.6)	2.59%	12.10% (3.48)	22.62%
Naive strategy (40 HFM samples)	<1% (0.2)	<1%	6.64% (1.20)	10.59%
MAC clustering (24 HFM samples, 6 clusters)	<1%(0.2)	<1%	6.72% (1.24)	12.09%
MAC clustering (24 HFM samples, 10 clusters)	<1% (0.2)	<1%	6.58% (1.11)	ss 9.56%

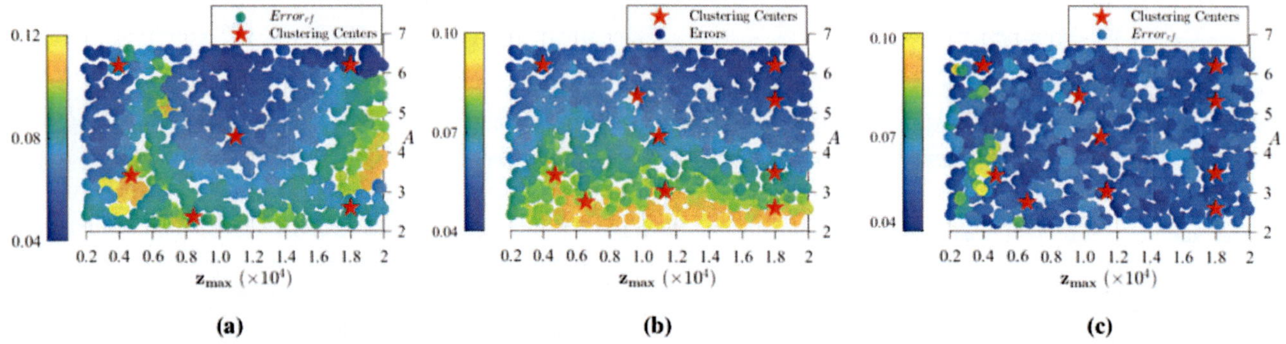

Fig. 4.3 Accuracy of the pROMs on capturing the nonlinear restoring forces. (**a**) Error with six clusters (Random Seed A). (**b**) Error with ten clusters (Random Seed B). (**c**) Naive sampling error

rate was adjusted to achieve similar performance with the MAC-based sampling. The respective error measures in Table 4.2 indicate that the MAC strategy delivers an accurate pROM approximation of the high fidelity dynamics without any prior knowledge of the input space. It also manages to do so with almost half the training samples, compared to the structured approach.

By increasing the number of clusters to ten the MAC-guided strategy delivers even better results using the same size of the training set. This is depicted in Fig. 4.3b. In this case, although the random seed of the initialization changed, the 6 out of ten clustering centers are located on similar regions as in Fig. 4.3a. Also, the rest four centers are placed in domains where the six clusters strategy of Fig. 4.3a experienced relatively low accuracy (eg $[1.8-2.0, 3-5]$). This correlation highlights the overall effectiveness of the MAC-guided strategy on indicating the sub-regions of the parametric space where the pROM is characterized by low precision to sample accordingly during training.

For instance, although the naive strategy has a lower maximum error in Fig. 4.3c compared to the six clusters in Fig. 4.3a, it seems to suffer from persistent high errors on the lower domain. As demonstrated in [24], this happens regardless of the sampling rate. However, the MAC-driven clustering in Fig. 4.3b captures this effect and samples in a finer rate in that region, reducing the error measures. This implies that such accuracy deterioration regions have been addressed explicitly, thus indicating an adaptive nature of the training process.

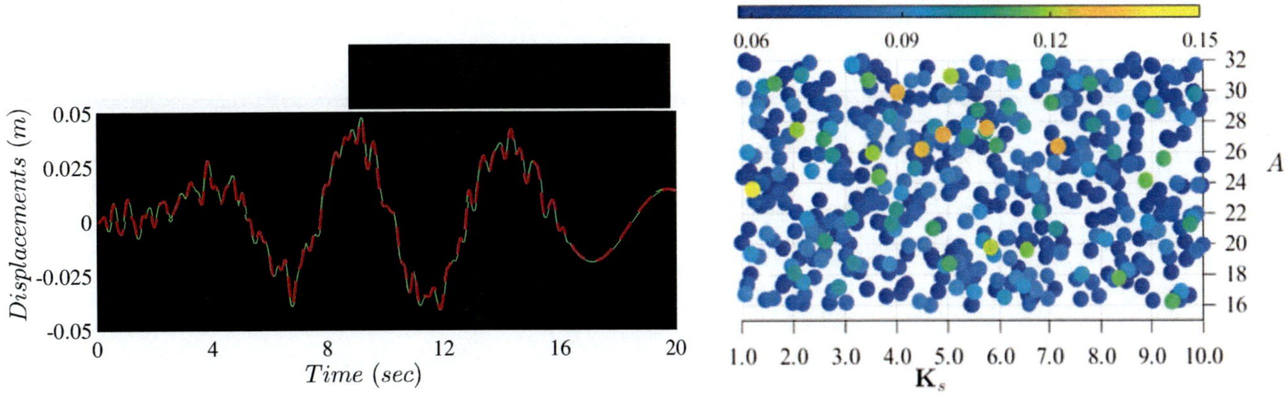

(a) (b)

Fig. 4.4 Response approximation and stresses error plot in 2D demonstrating the accuracy performance of the pROM. A coefficient of $\mathbf{K}_s = 10$ denotes a fully restrained basis. (**a**) Time history response approximation (Median Error 2%). The green line represents the full-order response whereas the red one the pROM approximation. (**b**) Maximum nodal stress error

4.3.2 Three-Dimensional Wind Turbine Tower with Parametrized Boundaries

This numerical case study derives a reduced representation approximating the numerical response of the NREL 5-MW baseline wind turbine tower [25]. A three-dimensional, shell finite element model is used, based on the detailed documentation in [24]. The Young modulus of the tower is 210 GPa and its density 7850 kg/m^3. The tower is assumed linear with a localized region characterized by isotropic von Mises plasticity spanning from 18% to 38% of the total height to test the substructural pROM formulation of this paper.

In the context of this study, the tower is excited assuming a $pi/4$ angle earthquake ground motion, approximated as a parametric low-pass filtered random signal. A white noise signal is passed through a low-pass filter, and the respective output is scaled based on an amplitude factor. The final processed signal represents the accelerogram of the ground motion. Thus, an amplitude parameter and the cut-off frequency are considered the model's parametric dependencies, similar to [24]. The earthquake excitation lasts for 10 s, whereas the model is simulated for 30 s.

Thus, dependencies pertaining to the earthquake motion's temporal and spectral characteristics are injected into the ROM. Also, the boundary conditions of the model are parametrized. Half of the nodes on the circular basis of the tower are considered fully constrained. The other half are modeled with linear springs, representing a loosening connection at every degree of freedom. The coefficient K_s of these springs is treated as an input parameter.

Based on the algorithm on Table 4.1, the input domain is sampled employing a lhs design and the MAC-based clustering initiates. After the pROM is trained, its performance is evaluated on a validation set spanning the whole domain. The high precision of the pROM in reproducing the time history response of the wind turbine tower is depicted in Fig. 4.4b for an example realization. As the Figure indicates, the approximation is almost indistinguishable, denoting an accurate reduced representation. Additionally, the pROM can capture the dynamic effects across the domain of parametric input, as demonstrated in Fig. 4.4b.

The ability of the derived parametric ROM to capture the underlying dynamics instead of simply reproducing the time history response is depicted in Fig. 4.5. The performance of the low-order representation is evaluated here concerning reproducing the stress state of the nonlinear component. Stresses and strains are fundamental quantities on structural health monitoring applications since they provide information on damage features. Thus, being able to approximate them is of utmost significance. Figure 4.5b presents a node-averaging stress visualization, similar to the high fidelity model in Fig. 4.5a. The respective maximum nodal stress error obtained is 14.5%, denoting a parametric ROM that can be employed for several applications that require precision on the strain or stress measures, including damage assessment or critical condition monitoring.

Regarding efficiency, the low-order representation achieved a maximum speed-up factor of 3, assuming both the high fidelity model and the ROM are evaluated on the same cores and using the same timestep during integration. However, this efficiency is strongly dependent on the nonlinear component's size and the high number of modes required for $\mathbf{V}_{z,q}$ ($r_x = 10$, $r_z = 50$ on Table 4.1). As long as the nonlinear component's size remains sufficiently small, compared against the linear domain, the efficiency due to the dimensionality reduction outweighs the evaluation of the nonlinear terms, thus leading to accelerated computations. However, in a more generalized case of multiple components or relatively large nonlinear

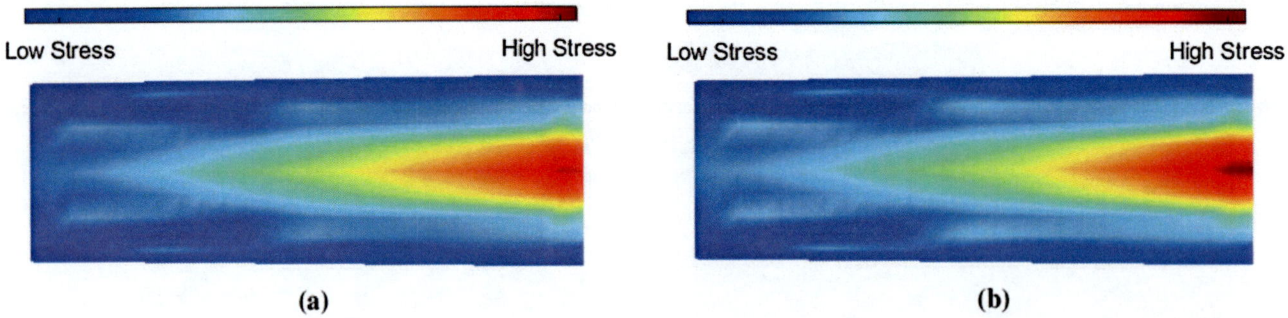

Fig. 4.5 Visualization of the the effective von Misses stress for the nonlinear component. (**a**) Stress visualization for the HFM. (**b**) Stress approximation of pROM

regions, hyper reduction methodologies are necessary to achieve a considerable computational toll reduction. To the authors' knowledge, the investigated case study is on the verge of these two scenarios. Thus, the speed-up factor may not be negligible but is small compared to the ones achieved by employing hyper reduction [24].

4.4 Conclusions

In this study, a physics-based parametric reduced order model was derived to capture response information from the high fidelity model in a twofold manner: From the global monolithic system dynamics, assumed as the ideal representation, and the localized components that experience different nonlinear effects. This was made possible by employing the substructuring formulation in [2, 19] that enables the individual treatment of components with localized features. The aforementioned substructural formulation of a parametric ROM is potentially critical in addressing systems with multiple nonlinear components and different in nature dynamics where traditional ROM methodologies either fail or require exhaustive sampling.

The performance of the parametric ROM was evaluated in a three-dimensional, real-scale case study featuring material nonlinearities. The pROM assumed no prior knowledge of the evolving dynamics across the parameter space. It utilized a clustering method based on the Modal Assurance Criterion to locate regions of similar dynamics and sample accordingly. The derived low-order representation captured the high fidelity system's dependencies pertaining both to the excitation and boundary conditions. The earthquake ground motion's temporal and spectral characteristics were parametrized, representing the uncertainty of the loading. Also, a loosening effect on boundary conditions was captured, representing the potential unfastening of model connections. The high precision on reproducing the time history response and the stress state of the model implies the potential of the pROM in several applications, including diagnostics and condition monitoring.

However, limitations of the presented approach do exist and formulate the orientation remarks for future research. Although the presented MAC-driven clustering captures the localized phenomena by correlating the POD bases, it requires at least a few full model evaluations. Thus, it cannot be used as an error indicator before sampling. Also, the MAC criterion does not guarantee a one to one mapping to the actual pROM error. In this study, it functions only as an indicator for regions with either similar or extremely different dynamics. Regarding computational efficiency, the implementation of hyper reduction techniques on the discussed framework should be considered, especially if the size of the nonlinear domain is relatively large or multiple nonlinear features are to be addressed.

Acknowledgments This research has been funded from the Sandia National Laboratories, the European Research Council, under the ERC Starting Grant WINDMIL (ERC-2015-StG 679843) on the topic of "Smart Monitoring, Inspection and Life-Cycle Assessment of Wind Turbines". Sandia National Laboratories is a multimission laboratory managed and operated by National Technology and Engineering Solutions of Sandia, LLC, a wholly owned subsidiary of Honeywell International Inc., for the U.S. Department of Energy's National Nuclear Security Administration under contract DE-NA0003525. This paper describes objective technical results and analysis. Any subjective views or opinions that might be expressed in the paper do not necessarily represent the views of the U.S. Department of Energy or the United States Government.

References

1. Benner, P., Gugercin, S., Willcox, K.: A survey of projection-based model reduction methods for parametric dynamical systems. SIAM Rev. **57**(4), 1–14 (2015)
2. Quinn, D.D.: Modal analysis of jointed structures. J. Sound Vib. **331**(1), 81–93 (2012)
3. Vlachas, K., Tatsis, K., Agathos, K., Brink, A.R., Chatzi, E.: A local basis approximation approach for nonlinear parametric model order reduction. Journal of Sound and Vibration **502**, 116055 (2021)
4. Klerk, D., de, Rixen, D.J., Voormeeren, S.: General framework for dynamic substructuring: history, review and classification of techniques. AIAA J. **46**(5), 1169–1181 (2008)
5. Krattiger, D., Wu, L., Zacharczuk, M., Buck, M., Kuether, R.J., Allen, M.S., Tiso, P., Brake, M.R.: Interface reduction for Hurty/Craig-Bampton substructured models: review and improvements. Mech. Syst. Signal Process. **114**, 579–603 (2019)
6. Allen, M.S., Rixen, D., Seijs, M., Van der, Tiso, P., Abrahamsson, T., Mayes, R.L.: Substructuring in Engineering Dynamics. Springer, Berlin (2020)
7. Tatsis, K., Dertimanis, V., Papadimitriou, C., Lourens, E., Chatzi, E.: A general substructure-based framework for input-state estimation using limited output measurements. Mech. Syst. Signal Process., 150 (2021)
8. Wu, L., Tiso, P., Tatsis, K., Chatzi, E., van Keulen, F.: A modal derivatives enhanced rubin substructuring method for geometrically nonlinear multibody systems. Multibody Syst. Dyn. **45**(1), 57–85 (2019)
9. Schoneman, J.D., Allen, M.S., Kuether, R.J.: Nonlinear modal substructuring of panel and stiffener assemblies via characteristic constraint modes. Dyn. Coupled Struct. **4**, 307–326 (2017)
10. Kerschen, G., Peeters, M., Golinval, J.-C., Vakakis, A.F.: Nonlinear normal modes, part I: A useful framework for the structural dynamicist. Mech. Syst. Signal Process. **23**(1), 170–194 (2009)
11. Kuether, R.J., Renson, L., Detroux, T., Grappasonni, C., Kerschen, G., Allen, M.S.: Nonlinear normal modes, modal interactions and isolated resonance curves. J. Sound Vib. **351**, 299–310 (2015)
12. Huang, X.-R., Jézéquel, L., Besset, S., Li, L., Sauvage, O.: Nonlinear modal synthesis for analyzing structures with a frictional interface using a generalized masing model. J. Sound Vib. **434**, 166–191 (2018)
13. Gruber, F.M., Rixen, D.J.: Dual Craig-Bampton component mode synthesis method for model order reduction of nonclassically damped linear systems. Mech. Syst. Signal Process. **111**, 678–698 (2018)
14. Lee, J.: A parametric reduced-order model using substructural mode selections and interpolation. Comput. Struct. **212**, 199–214 (2019)
15. Weeger, O., Wever, U., Simeon, B.: On the use of modal derivatives for nonlinear model order reduction. Int. J. Numer. Method. Eng. **108**(13), 1579–1602 (2016)
16. Joannin, C., Thouverez, F., Chouvion, B.: Reduced-order modelling using nonlinear modes and triple nonlinear modal synthesis. Comput. Struct. **203**, 18–33 (2018)
17. Im, S., Kim, E., Cho, M.: Reduction process based on proper orthogonal decomposition for dual formulation of dynamic substructures. Comput. Mech. **64**(5), 1237–1257 (2019)
18. Jin, Y., Lu, K., Huang, C., Hou, L., Chen, Y.: Nonlinear dynamic analysis of a complex dual rotor-bearing system based on a novel model reduction method. Appl. Math. Model. **75**, 553–571 (2019)
19. Quinn, D., Brink, A.: Global system reduction order modeling for localized feature inclusion. J. Vib. Acoust. **143**(4), 041006 (2021)
   Quinn, D., Brink, A.: Global system reduction order modeling for localized feature inclusion. J. Vib. Acoust. (2021). Submitted (under review)
20. Agathos, K., Bordas, S.P., Chatzi, E.: Parametrized reduced order modeling for cracked solids. Int. J. Numer. Methods Eng. **121**(20), 4537–4565 (2020)
21. Hesthaven, J.S., Stamm, B., Zhang, S.: Efficient greedy algorithms for high-dimensional parameter spaces with applications to empirical interpolation and reduced basis methods. ESAIM Math. Model. Numer. Anal. **48**(1), 259–283 (2014)
22. Allemang, R.J.: The modal assurance criterion–twenty years of use and abuse. Sound Vib. **37**(8), 14–23 (2003)
23. Simpson, T., Giagopoulos, D., Dertimanis, V., Chatzi, E.: On dynamic substructuring of systems with localised nonlinearities. In: Dynamic Substructures, vol. 4, pp. 105–116. Springer, Berlin (2020)
24. Vlachas, K., Tatsis, K., Agathos, K., Brink, A.R., Chatzi, E.: A physics-based, local POD basis approach for multi-parametric reduced order models. In: Conference Proceedings of ISMA-USD (2020). https://doi.org/10.3929/ethz-b-000442909
25. Jonkman, J., Butterfield, S., Musial, W., Scott, G.: Definition of a 5-MW Reference Wind Turbine for Offshore System Development. Technical report, National Renewable Energy Laboratory, Golden, Colorado (2009)

Chapter 5
On Predicting Uncertainties in the Dynamic Response of a Welded Structure

A. Muraleedharan, R. J. Barthorpe, and K. Worden

Abstract Present-day engineering projects are highly dependent on numerical models; thanks to the improvements in computing capabilities that have contributed significantly in this area. Although models these days are far better representations of engineering structures than before, every model is limited by its mathematical representation and the knowledge about the underlying physics. Validating numerical models involves obtaining test or performance data, which may not be practical in the case of many engineering structures. In such cases where data for the full structural model are not available, the subsystems or components can be tested and the associated models calibrated and validated separately. Inferring response at system level from these subsystem validation results is not straightforward and needs a proper uncertainty propagation strategy. Furthermore, the response of a structure depends not only on the components that it is made of, but also equally on the joints. Joints determine how the components interact within a structure, making validating the models for joints as important as validating the subcomponents. A joint does not physically exist on its own but co-exists with the components it connects, which makes it difficult to define a joint model. Isolated models, where the same type of joints is employed to connect 'simple' components with well-established numerical models can serve such a joint model for the purpose of validation. This paper describes a probabilistic approach to dealing with joints via such isolated models, where the uncertainties related to a welded joint are quantified and propagated into a target model of a welded structure to predict the dynamic response.

Keywords Joints · Welds · Validation · Calibration · Uncertainties · Probabilistic structural dynamics

5.1 Introduction

Recent decades have witnessed significant improvements in computational capabilities, which have had a direct influence on modelling engineering systems. In structural and mechanical engineering, better and more accurate representations of structures are possible as a result of this progress. This movement has also led to increased interest in understanding uncertainties in model predictions and methods to deal with them. The sources of uncertainty can be varying; this could be traced from the conceptual stage to the numerical solver and also include inherent uncertainties that exist in nature. Uncertainties can be broadly classified into *epistemic uncertainties* and *aleatory uncertainties*. 'Epistemic' refers to those uncertainties resulting from negligence and/or lack of knowledge of the process or the environment. They can be better defined as a potential inaccuracy in any phase or activity of the modelling process that is due to lack of knowledge [1]. These kinds of uncertainties can be largely reduced, if not eliminated, by expert knowledge and better modelling techniques, but this might come at an expense which may or may not be acceptable, depending on specific requirements. These uncertainties are also often biased to one side (either positive or negative). On the other hand, aleatory uncertainties are the inherent variation associated with the process or environment; they cannot be reduced and are often known as 'irreducible' uncertainties.

Having said that, it is essential to understand the various stages involved in the development of a computational model. All models are dependent on the knowledge of underlying physics. No knowledge is complete, and the concepts often ignore some non-trivial physics. Compromises in the model start from here, contributing directly towards epistemic uncertainties. The concept and understanding is then translated into a mathematical format which results in the mathematical model of a system. A concept without any mathematical backing is often useless in present-day engineering, and accurately representing

A. Muraleedharan (✉) · R. J. Barthorpe · K. Worden
Dynamics Research Group, Department of Mechanical Engineering, University of Sheffield, Sheffield, UK
e-mail: a.muraleedharan@sheffield.ac.uk; r.j.barthorpe@sheffield.ac.uk; K.worden@sheffield.ac.uk

M. S. Allen et al. (eds.), *Dynamic Substructures, Volume 4*, Conference Proceedings of the Society for Experimental Mechanics Series, https://doi.org/10.1007/978-3-030-75910-0_5

the desired concept by mathematics is often a challenge. In general, a mathematical model establishes the relation between different external conditions, various parameters that govern the system and the desired response quantities. These relations are often very general and need fine tuning or calibration to make them useful. Most engineering problems involve complex mathematical models such as partial differential equations or other complex functions, besides often being too big and impractical to solve analytically. It is at this stage that one may develop a numerical strategy to solve the mathematics, and this generally ends with a computer code. The numerical strategy might involve several discretisations and linearisations, applied to the mathematical model to enable an efficient computational framework. The predictions of the computational model should match reasonably accurately with the mathematical model, and necessary measures should be employed at this stage to make sure that the numerical model and the solver are reliable; this process is referred to as *verification*. Many problems in mechanical and structural engineering deal with finite element models, and discretisation errors are a major issue to be dealt with. Techniques like convergence analysis, *a posteriori* error estimation, and Richardson extrapolation [2] are widely used by designers and modellers to determine the solution approximation errors and account for them [3]. This application on a finite element problem is demonstrated in [4]. Surrogate models (also called as meta-models) are also widely used in practice today, and the modelling errors associated with them depend on the type of surrogate models.

A verified model ensures its capability to solve the underlying mathematics with acceptable accuracy, but it is by a calibration exercise that one can fine tune the mathematical equations to achieve predictions that match observations in practice. Calibration fine tunes the model to predict observed data, but the reliability of this model needs to be validated. Validation directly answers the question of whether the model is right, i.e. whether the conceptual model (and our understanding of the physics) is right. The process of calibration and validation are explained in the following sections in more detail. For both calibration and validation operations, experimental or performance data are needed. Engineering structures are often complex enough to make it practically hard to obtain the necessary data in a test facility. This limitation is overcome to an extent by subsystem-level calibration and validation. While the structure in its entirety remains hard to test, different parts of the same structure might be possible to test. The models of these parts or substructures referred to as subsystem-level models could thus be calibrated and validated. These calibrations of subsystem-level models are often directly helpful in updating relevant parameters in a model representing the full structure. Furthermore, if the full structure is modelled by integrating the validated subsystem-level models, it can further improve the quality and reliability. By these subsystem-level operations, the uncertainties regarding the parameters can also be quantified using an appropriate calibration strategy such as a Bayesian approach [5]. This understanding of the uncertainties at the subsystem level also facilitates computation of uncertainties in the predictions at the system level.

While this idea remains dependable, it has to be taken into account that a structure is not composed of just substructures or parts, but also of the joints which connect these parts as intended. The joints determine the interaction between the parts and hence are as important as the parts themselves. Joints often do not get the deserved attention in structural models, as they are not easy to model. This issue is partly due to the lack of in-depth knowledge about the underlying physics and partly due to the complexities in representing and implementing the physics into a model. Even in the case of a simple joint model, calibration of joint-specific parameters is not easy. A joint does not exist on its own, and this makes it difficult to obtain any relevant information regarding the joint parameters, especially when the parts are tested individually without the joints themselves. The uncertainties, if quantified, result in a better system-level model.

This paper focuses on joints and ways to address them with the same importance as any other part in a structure. A concept of addressing joints as separate subsystems and performing the calibration and validation operations is discussed. This concept is detailed through an example problem involving a welded joint and the subsequent calibration and validation operations specific to that joint.

5.2 Dealing with Joints

5.2.1 Isolated Joint Model

The difficulty in properly characterising a joint often leads to unaccounted uncertainties in system-level predictions. The practical difficulty in calibrating joint-specific parameters exists in cases where the subsystems are tested in isolation. Even in cases where some of the joined structures are tested, often the uncertainties regarding the joint-specific characters are blended with the uncertainties in structural properties of the connected components, which makes it difficult to identify each accurately. In this context, if the joint could be treated as a separate subsystem with its own subsystem-level models, this could manage the situation better. A prospective idea here is to depend on isolated joint models, which are specifically designed to identify and calibrate various parameters that govern a joint.

Some parameters governing a joint might be defined in general terms, but many of the joint-specific parameters can only be observed on a joint in action, or only when the joint exists in the desired configuration. In this regard, the proposed isolated joint models and the corresponding test specimens are connected components, where two well-defined components are connected by the very same type of joint in a specified configuration. The advantage here is that each of the connected components would have well-defined models associated with them, which are calibrated and validated. This approach reduces the problem to a condition where the uncertainties regarding the components are minimised and any existing uncertainties are quantifiable. In such a case, the parameters governing the joints can be calibrated with suitable test data with more ease. The calibration exercise regarding this isolated system is essentially the calibration of joint-specific parameters. Further validation of this model strengthens confidence in the model, i.e. regarding the joint model or the joining concept and physics that forms the basis of the model. Such a model, or essentially the concept or validated physics of the joint, can be employed in any system-level model with similar joints with the calibrated joint-specific parameters. The uncertainties in the parameters are also captured via the calibration exercise. This concept could be better understood via a simple example. If a structure has an associated model at system level H, let the mathematical model be represented as,

$$Y_S = H(X_S, \theta_S) \tag{5.1}$$

where Y_S is a response predicted by the model to an input X_S and θ_S represents the parameters that define the model. If this structure is formed by two substructures (or subsystems) S_1 and S_2, and G_1 and G_2 are the corresponding subsystem-level models, their mathematical models can be represented as,

$$\begin{aligned} Y_{S_1} &= G_1(X_{S_1}, \theta_1) \\ Y_{S_2} &= G_2(X_{S_2}, \theta_2) \end{aligned} \tag{5.2}$$

In this situation, θ_1 and θ_2 are subsets of θ_S, and θ_S may have other parameters which are not addressed by either of the subsystems, such as those parameters specific to the joints θ_J. Calibrating and validating the subsystem-level models enables one to feed the system-level model H with the updated values of the parameters θ_1 and θ_2 with any quantified uncertainties. However, an isolated joint model needs to be designed, tested, calibrated and validated to update θ_J and to quantify any associated uncertainties. Since one can assume a joint is a separate subsystem, this approach can be achieved by a joint model J, where the corresponding isolated system is one where two well-defined structures are connected,

$$Y_J = J(X_J, \theta_J, \theta_0) \tag{5.3}$$

where θ_0 refers to the parameters defining models of the connected well-defined components in the isolated system which are calibrated. On performing calibration and validation on this model, one can update θ_J with the associated uncertainties quantified and use it in model H to obtain better results. It has to be taken into consideration that the model H might possess additional parameters which are not all dealt with by either of the subsystem-level models or the isolated joint model. This point remains a source of further unidentified uncertainties; nevertheless the uncertainties regarding the joint and the system as a whole could be estimated to a better degree by adopting these methods.

The isolated joint models should be selected carefully to address most of the significant joint-specific parameters in the system-level model under consideration. In this regard, even multiple isolated joint models can be employed if a single isolated model may fail to capture any significant parameters. Using multiple (different) isolated joint models also improves the confidence and reliability in the calibrated parameters. The paper [3] describes a method to integrate inferences from multiple models at lower levels.

5.2.2 *Calibration*

Calibration refers to the process of fine tuning of parameters in a model by comparing the results of predictions with experimental data. When one looks back at the mathematical model, it is these parameters that give a form to the mathematical equations, by defining various relations between the input and responses. Most of the parameters in structural problems have some form of associated physical meaning. In this paper, the calibration is achieved via a probabilistic approach relying on the concepts of Bayesian statistics [3, 5]. In such an approach, the initially assumed values for the parameters (called priors) are updated based on the likelihood (the probability of obtaining the test data with the given parameters) to obtain the posteriors (updated values of the parameters). If $p(\theta_J)$ is the prior distribution of the parameter θ_J of a model J with calibration data D^c, the posterior of the parameter $p(\theta|D^c)$ is given by,

$$p(\theta_J|D^c, J) = \frac{P(D^c|\theta_J, J)p(\theta_J)}{\int P(D^c|\theta_J, J)p(\theta_J)d\theta_J} \tag{5.4}$$

$P(D^c|\theta_J, J)$ is the likelihood function of θ_J and assuming a normally distributed likelihood function, it can be expressed as,

$$P(D^c|\theta_J, J) \propto \prod_{i=1}^{n} \frac{1}{\sigma\sqrt{2\pi}} e^{-\frac{(Y_i - J(X_i, \theta_i)^2)}{2\sigma^2}} \tag{5.5}$$

where σ is the standard deviation of the observation errors.

5.2.3 *Validation*

Model validation refers to the process of determining the degree to which a model (and its associated data) is an accurate representation of the real world from the perspective of the intended use of the model [6]. Defining this in more practical terms, validation is the process of comparing model results and their associated uncertainties, with experimental results and their associated uncertainties [7]. Validation of models is crucial in establishing the reliability of predictions. There are different 'validation metrics' to describe the degree of validation qualitatively and quantitatively [8–11]. In general, these metrics can be classified into two main categories—Frequentist and Bayesian. In this paper, the *area metric* proposed by Ferson et al. [12] is used as a validation metric. This popular method measures the area between two *cumulative distribution functions* (CDFs), in this context, between CDFs of predictions and observations. The area is computed as the Minkowski L_1 metric. If $F_p(x)$ and $F_m(x)$ are the CDFs of predictions and observations, respectively, the area metric is given as,

$$D_{area}(p, m) = \int |F_p(x) - F_m(x)| \tag{5.6}$$

The Bayes factor B [13] can also be used as a validation metric. Bayes factor is defined as the ratio of likelihood of the null and alternate hypothesis, which could be the likelihood ratio of two competing models. If two models J_1 and J_2 are competing against each other and D^v is the observed or measured validation data, then

$$B = \frac{P(D^v|J_1)}{P(D^v|J_2)} = \frac{\frac{P(J_1|D^v)p(D^v)}{P(J_1)}}{\frac{P(J_2|D^v)p(D^v)}{P(J_2)}} = \frac{P(J_1|D^v)P(J_2)}{P(J_2|D^v)P(J_1)} \tag{5.7}$$

A value of Bayes factor B larger than unity favours the model J_1, and a value less than unity favours the model J_2. From B, the respective probabilities of each model to be selected can also be obtained,

$$P(J_1|D^v) = \frac{B}{B+1} \tag{5.8}$$

$$P(J_2|D^v) = 1 - P(J_1|D^v) = \frac{1}{B+1} \tag{5.9}$$

In a situation where the validity of a single model J_1 is considered (i.e. without any competing models) the model J_2 can be replaced with a suitable alternate hypothesis J_1' [3, 13]. In the latter part of the paper, the Bayes factor is used to compare the performance of calibrated and uncalibrated models.

5.3 Welded-Joint Model

5.3.1 The Weld as a Joint

As discussed earlier, the purpose of a joint is to connect two parts together, enabling them to perform as a single entity in the desired configuration, or to realise a desired interaction between the connected parts. In a welded joint, the connection is achieved by introducing a continuity between the parts by fusing the materials at the joint [14, 15]. This property distinguishes a weld from other types of joints. The fusion is generally achieved by melting the materials at the interface and allowing them to mix and cool down [16]. Often, a filler material is employed, which introduces extra material into the joint to make the process easier. Note that, the discussion in this paper is limited to metal structures. There exist different methods of achieving a weld, depending on the geometry, material properties, environment and requirements.

The type of weld assumed here is a generic butt weld, as illustrated in Fig. 5.1. A butt weld is intended to achieve continuity of the material, so that the overall structure replicates as closely as possible, a monolithic structure. The joining components are prepared and aligned before introducing an electric arc (or other suitable method) to melt the metal at the region, while the filler material (also molten) is also introduced. The filler material fuses with the parent material at the interface. The heat can also cause molecular-level changes around the welded area; these regions being usually referred to as *heat affected zones* (HAZ). The reference [14] covers in detail, the process of welding. From an engineering point of view, a weld in a structure introduces a material and structural discontinuity. It is an obvious observation that a welded structural member is not the same as a monolithic cast member. However, it is often difficult to account for this discontinuity at the welded joints. The structural properties in the region have significant uncertainties, which make it difficult to address these welded regions perfectly in a design. If the uncertainties pertaining to the weld parameters could be identified, it would improve the predictions of models of structures involving welds.

5.3.2 Model Description

The purpose of the model discussed here is to predict the dynamic response [17] of a welded beam. In general, models are calibrated via experimental observations; what makes this difficult in some complex structures is the difficulty in performing tests on the system as discussed in previous sections. In this context, employing isolated models, as discussed in Sect. 5.2, to calibrate and validate a reliable model for the welded joint would enable designers to improve models involving welds. The isolated model considered here is a welded cantilever beam, where two identical sections of a beam are joined using a butt weld. Here, the individual beams are assumed to have well-defined models (verified, calibrated and validated); this enables one to identify the uncertainties in parameters governing the individual beams. Hence, any additional uncertainties in the predictions for the welded beam point to uncertainties from the 'weld' and the parameters that define the weld. This approach is a form of divide-and-solve approach, where, instead of calibrating all the parameters of the welded beam on its own, the available models of individual beams are already calibrated and validated in advance; the problem is thus crafted to concentrate on the specific area of interest, where one can specifically update the parameters governing the weld. This approach also helps reduce the dimension of the calibration problem, resulting in better accuracy in updated parameters compared to updating them simultaneously.

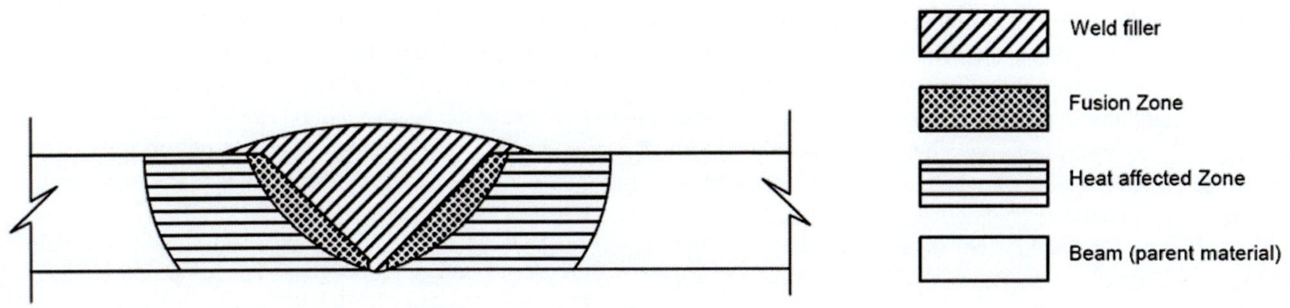

Fig. 5.1 Details of a typical butt weld

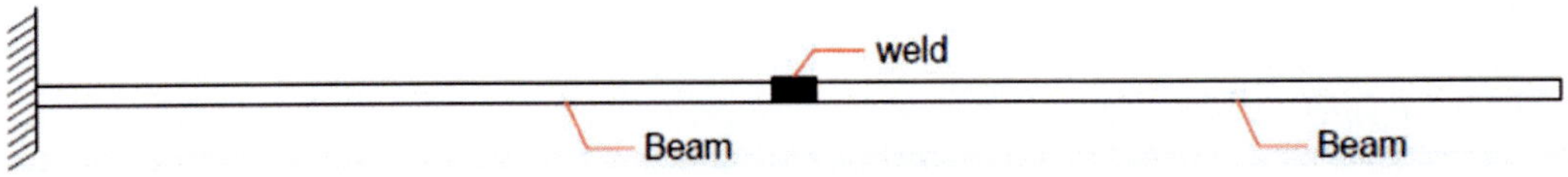

Fig. 5.2 Representation of the welded beam

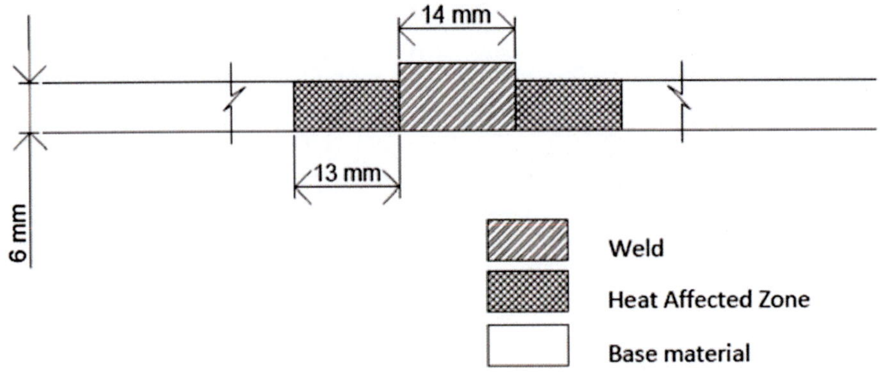

Fig. 5.3 Simplified model of the weld region

As this paper is intended to present a concept, and, not to accurately represent a physical weld, a simpler representation of a welded beam model is employed. Also, the observations are numerically simulated using a set model with initially assumed parameters. The parameter values used to synthesise the responses are referred to in the paper as the 'true parameters' and these are given in Table 5.2. The welded beam assumed here is illustrated in Fig. 5.2, where the different regions are marked. The welded region is also represented as shown in the Fig. 5.2. The beam is modelled using a finite element approach with Euler–Bernoulli beam elements [18]. This approach derives from the assumption that the welded structure is bonded by the weld material and that a perfect weld is free of any voids or cracks.

The overall length of the welded beam under consideration is 500 mm. At the weld, one can observe different regions based on the material and thermal effects of the welded joint and the welding process. The different regions can be broadly classified as those shown in Fig. 5.1, namely the parent material, heat affected zone, fusion zone and the weld material. The fusion zone is generally a very thin layer forming the boundary between the weld material and the HAZ, whereas the HAZ are those regions which were exposed to high temperatures during the welding process. The fusion zone is neglected in this paper. Also to simplify the model, the weld and heat affected zones are assumed as illustrated in Fig. 5.3. In the finite element model, each element has a length (mesh sizing) of 20 mm. HAZ (one element each) and the welded region (2 elements) are also modelled with the same approach but considering the differences in material properties and the thickness at the respective regions. The dynamic responses (displacements) are computed via a state-space approach. All the modelling and computations are performed using Matlab. The important parameters here are the shape and size of the structure, as well as the material properties. The problem is assumed to be linear, and the beam is considered to be welded perfectly to avoid any geometric distortion. Therefore, the factors affecting the dynamics are limited to the geometry, mass, stiffness and damping parameters.

The parameters that are specific to the welded joint here include: (1) material level stiffness of the weld itself and the adjoining HAZ to either side of the weld, (2) damping at the weld and the HAZ and (3) the increased thickness at the weld which contributes to mass and stiffness. The density is assumed to be constant throughout each region and deterministic. Furthermore, the stiffness at the heat affected zone does not change significantly due to the heating process, and therefore the Young's modulus 'E_{HAZ}' of the HAZ is considered to be the same as that of the beam section. The Young's modulus of the weld E_{weld} is a parameter which needs to be calibrated in this problem. E_{weld} is assumed to have a uniform prior and assumed to possess a value anywhere within a given range. Another important parameter to be calibrated is the damping in the system. The parameters that define the structure and their respective values are summarised in Table 5.1. The randomness in the thickness of the weld is considered via a normally distributed variance; this parameter is assumed to be measurable and is not calibrated.

Although damping [19] in metals is very low, it is of interest to see its effect because of a welded joint. Experiments to measure damping are necessary to understand the magnitude of the damping and effect of the weld on damping, and thus this influences the overall dynamics. The actual effect of damping in a welded structure might be better understood via a real experiment (this might influence the decision to consider the damping parameter for calibration); however, this paper aims to

Table 5.1 Parameters governing welded beam

Parameter	Description	Value
L	Total length	500 mm
L_{weld}	Extend of weld	14 mm
L_{HAZ}	Extend of HAZ	13 mm
t_{beam}/t_{HAZ}	Thickness of beam and HAZ	6 mm
t_{weld}	Thickness of weld	$(\mu = 8, \sigma = 0.2)$ mm
E_{beam}/E_{HAZ}	Young's modulus of beam and HAZ	$(\mu = 1.9, \sigma = 0.01)$ GPa
E_{weld}	Young's modulus at weld (uniform distribution)	[1.6, 2.4] GPa
ν	Poisson's ratio	0.3
ρ	Density	$2700\,\mathrm{kg/m^3}$
η_{beam}	Damping loss factor—beam	$(\mu = 0.004, \sigma = 0.0002)$ Ns/m
η_{weld}	Damping loss factor—weld region (uniform distribution)	[0.05, 0.07] Ns/m

Table 5.2 True values of parameters used to synthesise observations

Parameter	Description	Mean	Standard deviation
E_{weld}	Young's modulus at weld (GPa)	2	0.1
η_{weld}	Damping loss factor: weld region (Ns/m)	0.06	0.002

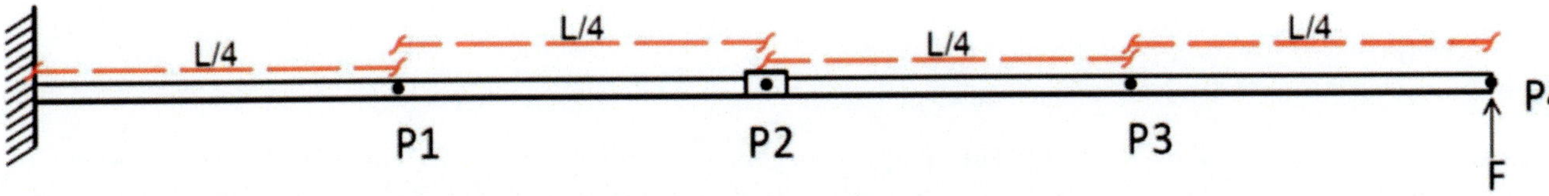

Fig. 5.4 Location of response measurement points on the model beam

create a theoretical framework to deal with the situation. Hence, the problem will proceed with the assumption that damping plays a role in the overall dynamics and that it needs to be calibrated. It is also assumed that the welded region exhibits a higher damping than that of the beam. The HAZ would also affect the damping characteristics in the region; being exposed to high temperatures can make molecular changes which affects the damping behaviour. In this problem, the damping is classified into two specific regions: the damping related to the base material and effective damping due to the welded region which includes the weld with the adjacent HAZs. While the material loss factor η_{beam} (which determines damping [19, 20]) for the beam is assumed to be known, the effective loss factor at the welded region η_{weld} is assumed to be unknown; i.e. this parameter needs to be calibrated via observed data.

5.4 Calibration

The parameters to be calibrated in this problem are the Young's modulus at the weld E_{weld} and the effective damping at the weld region η_{weld} (this region includes the weld and HAZ). Calibration data are simulated from the model using the true values of the parameters as shown in Table 5.2. The model of the isolated welded beam J computes responses, which are time-domain displacement responses at four points $P1$ to $P4$ as illustrated in Fig. 5.4, resulting from an impulse force of 100 N applied at the point $P4$. Responses were recorded at 1000 Hz resolution for a duration of 5 s. These points are located at each quarter of the length of the beam. To mimic a real observation a white noise (with RMS 4×10^{-5}, which is around 1% of the RMS of the displacement response time histories at $P4$) is added to the simulated responses. From these responses, the natural frequencies and associated damping ratios were identified (for the first two modes), using a stochastic subspace identification (SSI) algorithm [21, 22].

The calibration problem involves two parameters, the Young's modulus of the weld E_{weld} and the effective damping at the weld region η_{weld}. These are estimated via a Bayesian inverse problem, where a Transitional Markov chain Monte Carlo (TMCMC) [23, 24] algorithm is used to determine the posteriors of both the parameters. This analysis is performed in two steps, as the two parameters are solved via different forward models. To update the Young's modulus of the weld, E_{weld},

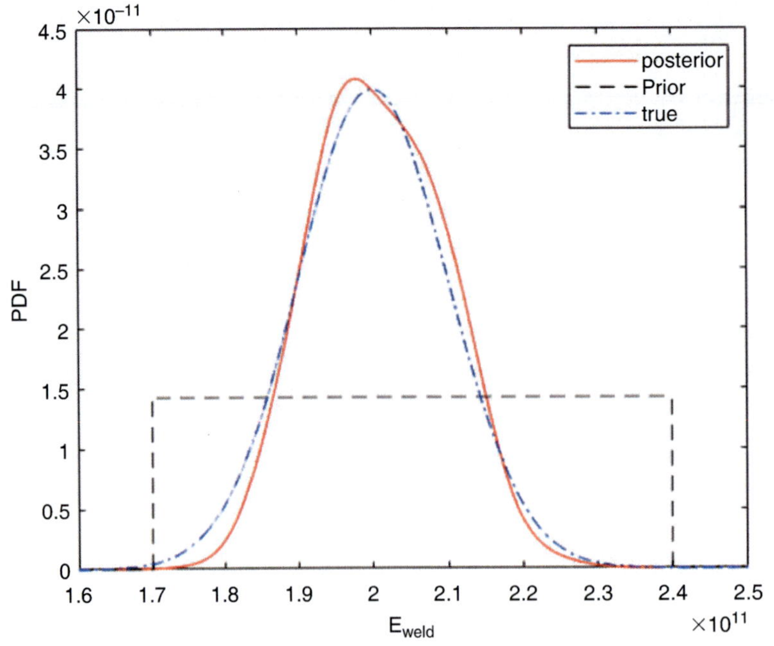

Fig. 5.5 Probability distribution of E_{weld}—posterior, prior and true

the observations considered are the observed natural frequencies (obtained by performing SSI on simulated observations). For the purpose of Bayesian calibration, the forward problem (forward-model 1) computes the same natural frequencies by solving the eigen value problem after developing the mass and stiffness matrices. Forward-model 1 refers to this model which computes the natural frequencies by solving the eigen value problem. Inverse Bayesian analysis is then employed to update for E_{weld}. Once E_{weld} is calibrated via this model, it is propagated into the FE model of the beam to determine the response time histories and subsequent damping ratios through forward-model 2. Forward-model 2 computes the response time histories with the updated values of E_{weld} and then computes the modal damping ratios through the SSI algorithm. The damping parameter, i.e. the effective loss factor at the weld region η_{weld} can be calibrated for this model, via another inversion problem.

The results of calibration of E_{weld} are illustrated in Fig. 5.5. The red line represents the posterior (updated) distribution for E_{weld}, while the black dashed line represents the prior. The blue dash-dot line refers to the true distribution of E_{weld} employed to simulate the observations. This updated distribution of E_{weld} is then propagated through forward-model 2 to predict the responses and the damping ratios. This process enables one to calibrate the damping parameter, i.e. the effective loss factor at the weld region. The results of the calibration are presented in Fig. 5.6. As in Fig. 5.5, the red line represents the posterior values of η_{weld} while the blue dash-dot line refers to the original distribution used to simulate observations. The black dashed line is the assumed prior for η_{weld}. The mean and standard deviation of the updated parameters are given in Table 5.3. The updated parameters show good agreement with the true values in Table 5.2. The parameters related to the weld being calibrated may be used to predict responses in structures with similar welds. Before employing these values in other models, one also has to confirm that the isolated joint model J is validated.

5.5 Validation and Comparison

For isolated joint model J being calibrated, it is essential to make sure that the calibrated model predicts the response close to the true responses. A completely different response from that considered for the calibration is being used here for validating the model. The responses at the end of the beam—point $P4$ in Fig. 5.4—for a periodic loading at point $P1$ (quarter the length from the support) of the beam are considered. The measurements are numerically simulated from the original beam with the exact parameters. In order to mimic experiment, a white noise of 0.5% of the RMS of the signal is added to the responses. The force applied is sinusoidal with an amplitude of 100 N and a frequency of 100 Hz.

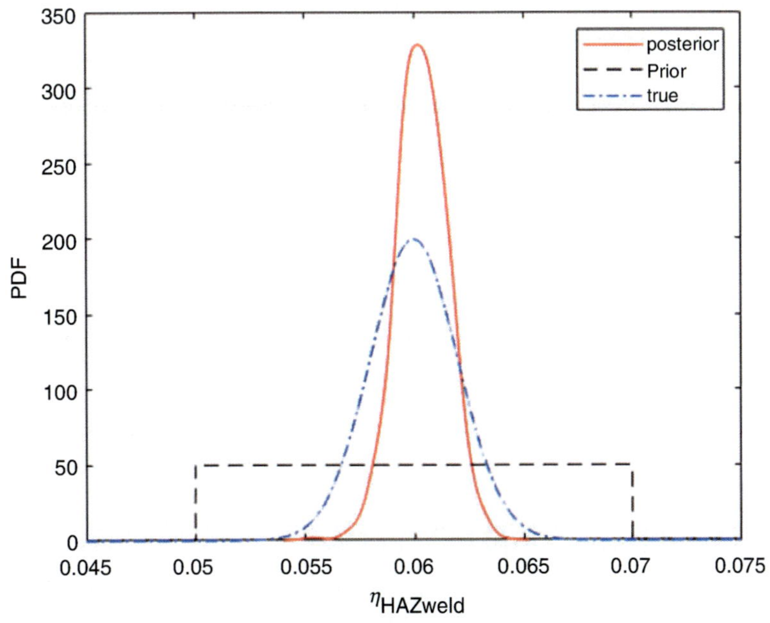

Fig. 5.6 Probability distribution of η_{weld}—posterior, prior and true

Table 5.3 Calibrated values of parameters

Parameter	Mean	Standard deviation
E_{weld}—Young's modulus at weld (GPa)	2.0072	0.083365
η_{weld}—Damping loss factor—weld region (Ns/m)	0.06	0.0011

The responses are obtained for a duration of 30 s at 1000 Hz resolution, of which the initial 15 s were eliminated to remove transient response. The same displacement time histories at the end were also obtained from the model under consideration (model J). This response is in the form of a vector representing the displacement at each time instance (a time series). As validation involves comparing the responses, the responses predicted from the model $[Y_p]^{(1\times T)}$ need to be compared with the responses measured $[Y_m]^{(n\times T)}$; T represents the number of points in the time series and n refers to the number of observations made. Here, n sets of observations are made, each under the same conditions and input loading. Since there are n sets of responses, the mean of these n responses is considered as the reference response $[Y_R]^{(1\times T)}$. Let $[Y_R] = [y_{R1}, y_{R2} \dots y_{RT}]$ and the ith measured response be $[Y_m^i] = [y_{m1}^i, y_{m2}^i, \dots y_{mT}^i]$, then,

$$y_{Rt} = \frac{1}{n}\sum_{i=1}^{n} y_{mt}^i \tag{5.10}$$

Once the reference response matrix $[Y_R]$ is calculated, the predicted response $[Y_p] = [y_{p1}, y_{p2} \dots y_{pT}]$ can be directly compared by computing the difference between the responses at each time interval. Errors in prediction at each time instant e_p^t can be calculated as,

$$e_p^t = y_{Rt} - y_{pt} \tag{5.11}$$

The sum of absolute values of errors (averaged over the length of time vector T) then gives a scalar estimate to compare between the two responses ($[Y_R]$ and $[Y_p]$). This measure is represented as E_p,

$$E_p^t = \frac{1}{T}\sum_{t=1}^{T} |(e_p^t)| \tag{5.12}$$

In a deterministic approach, E_p will be a single value, but considering the uncertainties in the model predictions, E_p is a random quantity; thus one can represent E_p with a probability distribution. The PDFs of E_p can be obtained from the model

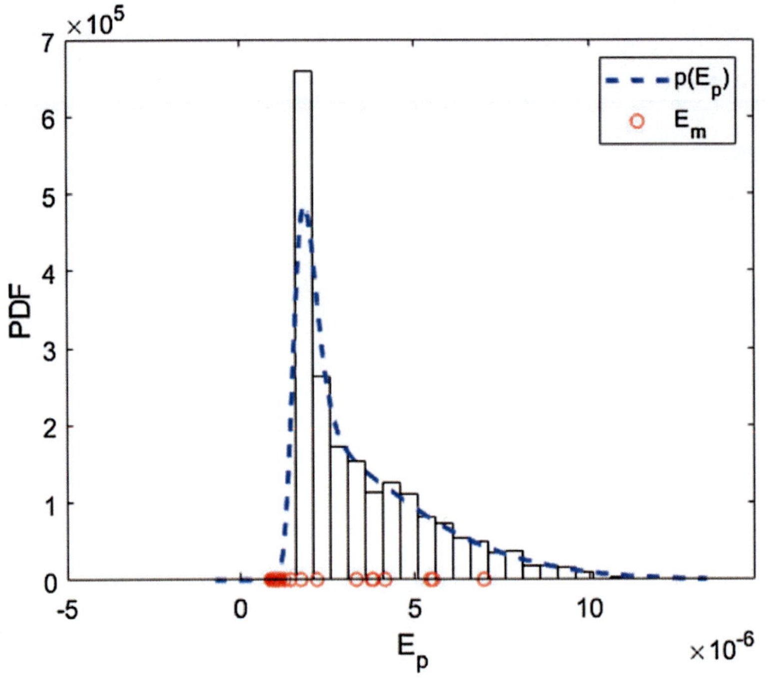

Fig. 5.7 Histogram of predictions and $p(E_p)$ compared to measurements E_m and the distribution $p(E_m)$

J, after propagating the uncertainties in the parameters through the model J; this is done by generating samples from the updated distribution (posteriors) of the parameters and obtaining the responses. The calibrated parameters were fitted with a kernel density estimate in Matlab; samples can also be generated from the fitted distribution function. This probability distribution takes into account the uncertainties in observations and gives an idea about the variability of the observations. The histogram of E_p as a result of the Monte Carlo simulation is represented in Fig. 5.7, where the probability distribution $p(E_p)$ is given as the blue-dashed line. It has to be noted that, as E_p is computed as the sum of absolute values of errors, there cannot be any negative values. Similarly, the measurement errors e_m^t and the sum of absolute values of the measurement errors (averaged) E_{mi} (red in Fig. 5.7) are also computed for all n sets of observations to obtain $[E_m]^{(n\times 1)} = [E_{m1}, E_{m2} \ldots E_{mn}]$.

The process of validation involves comparing the predictions with the measurements; here, the comparison is between the prediction errors to the measurement errors. The validation metric being used here is the *area metric* (5.2.3) which computes the difference between two distributions. The area between CDFs of prediction $F(E_p)$ and measurement $F(E_m)$ is computed here to obtain a metric which directly relates to the differences between them. The CDFs are illustrated in Fig. 5.8; for the measurement errors E_m, CDF refers to an empirical CDF [12, 25]. The area metric, geometrically, is the area between the two CDFs (between the red and blue-dashed lines in Fig. 5.8); a value of zero indicates the ideal case of a perfect model with exact match to the observations. This method also possesses the advantage of geometrically visualising the deviation from measurements. The area metric calculated using the formula given in Eq. (5.6) is 1.47×10^{-6} m. In engineering practice, there could be thresholds respecting the requirements to accept a model based on the area metrics. Here, from Fig. 5.8 and from the value of area metric, it is evident that the calibrated model does a good job in predicting the responses.

Once the isolated joint model J is validated, the calibrated parameters defining the weld can be propagated through a welded joint modelled in another structure with similar welds. This process enables one to make better predictions regarding the new structure which may not be possible to calibrate by itself in the joined (welded) configuration. It should be noted that care is needed about any additional joint parameters that might be specific to the structure which might impose uncertainties in the predictions.

Although the model has been validated using an area metric, it is of obvious interest to compare the calibrated model with the calibrated model. It is quite natural to assume the prior values for the joint parameters when isolated joint models are not employed. Here, in order to compare the two models, a Bayes factor is utilised. As an alternate model to the model J, the same model without calibrating the weld is considered, or in other words, the weld parameters assume the prior values. This alternate model is referred to as J' and predictions Y_p' from this model are also compared with the reference response $[Y_R]$ to get the sum of absolute values of errors E_p'. Assuming no initial knowledge, both the models are considered to be equally probable in predicting the observed responses; in other words, the priors are equal. The probability distribution $p(E_p')$ is

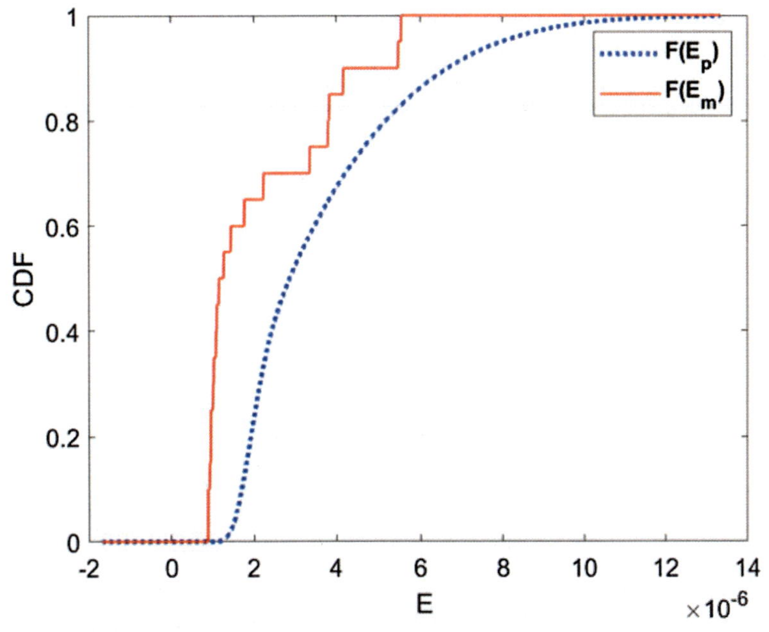

Fig. 5.8 Cumulative distribution functions for predictions $F(E_p)$ and measurements $F(E_m)$

also obtained in the same manner as in the previous case, using a Monte Carlo approach. Instead of considering E_m as a vector, the probability distribution $p(E_{mi})$ is obtained. E_m is fit with a Gamma distribution (as the absolute values of errors are positive). Now that the probability distributions $p(E_m)$, $p(E_p)$ and $p(E_p')$ are obtained, one can estimate the likelihood of both of the models, based on the measured responses $[Y_m]$. The likelihoods of both models, J and J' are given as,

$$L(J) = P(E_m|J) = \prod_{t=1}^{n} P(E_{mi}|J) \tag{5.13}$$

$$L(J') = P(E_m|J') = \prod_{t=1}^{n} P(E_{mi}|J') \tag{5.14}$$

where $L(J)$ and $L(J')$ refer to the likelihoods of models J and J', respectively. The Bayes factor, which is the ratio of the likelihoods between the null and the alternate models, is utilised for comparing both the models.

$$B = \frac{L(J)}{L(J')} \tag{5.15}$$

Figure 5.9 compares $p(E_m)$ (red) with $p(E_p)$(blue dotted) and $p(E_p')$ (in black dash-dot); the red circles refer to the measured errors E_{mi}, and one can clearly observe the higher chances of obtaining a greater error in model J'.The likelihood (probability of obtaining the measured distribution from the model) can be calculated from,

$$L(J) = p(E_m|J) = \int_{E=0}^{\infty} p(E_m) \times p(E_p)dE \tag{5.16}$$

$$L(J') = p(E_m|J') = \int_{E=0}^{\infty} p(E_m) \times p(E_p')dE \tag{5.17}$$

The Bayes factor can be calculated as the ratio of the likelihoods as described before; a factor greater than 1 favours the model J and a value less than one favours the model J'. As evident from the Fig. 5.4, a very high Bayes factor $B = 7.94$ is obtained which suggest that the model J is preferred to the model J'. These results illustrate the importance of calibrating the weld in a structure.

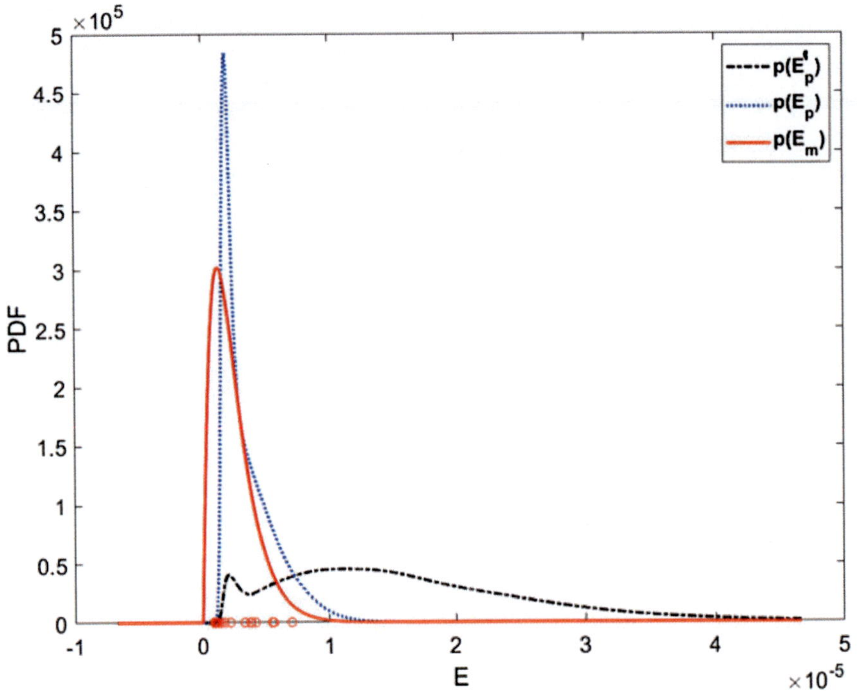

Fig. 5.9 Probability distributions of $p(E_p)$, $p(E'_p)$ and $p(E_m)$

5.6 Discussion and Conclusions

Considering joints, joint-specific parameters in a structural model are often associated with greater uncertainties than the connecting components. Often in complex structures, the structure cannot be tested in the assembled configuration to calibrate and validate the system-level models. While calibrating via subsystem-level models, joints are often neglected and prior values of parameters are assumed depending completely on human knowledge and engineering judgement. An idea to consider the joint as a separate subsystem via a specially selected joint model and corresponding test specimen has been discussed here. This method opens the possibility of calibrating for significant joint-specific parameters, which can be directly employed in a system-level model after validation. The use of probabilistic methods was also demonstrated via a Bayesian framework, which could capture the uncertainties regarding the parameters which directly contribute to uncertainties in predictions. This concept has been demonstrated with the example of a welded joint where the isolated joint model of the weld was calibrated and validated. The authors are working on extending this approach to a real welded joint with the support of laboratory experiments and measured data to identify and calibrate dominant parameters that govern a weld. Future works are expected to develop and demonstrate more clearly various strategies regarding validation models regarding welded structures.

Acknowledgments This project is supported by funding from the EU's Horizon 2020 research and innovation programme under the Marie Sklodowska-Curie grant agreement number 764547 (DyVirt ETN).

References

1. Oberkampf, W.L., DeLand, S.M., Rutherford, B.M., Diegert, K.V., Alvin, K.F.: Error and uncertainty in modeling and simulation. Reliab. Eng. Syst. Saf. **75**(3), 333–357 (2002)
2. Richards, S.A.: Completed Richardson extrapolation in space and time. Commun. Numer. Methods Eng. **13**(7), 573–582 (1997)
3. Sankararaman, S., Mahadevan, S.: Integration of model verification, validation, and calibration for uncertainty quantification in engineering systems. Reliab. Eng. Syst. Saf. **138**, 194–209 (2015)
4. Mestrovic, M.: An application of Richardson extrapolation on FEM solutions. Int. J. Math. Comput. Method. **1**, 351–354 (2016)
5. Kennedy, M.C., O'Hagan, A.: Bayesian calibration of computer models. J. R. Stat. Soc. Ser. B (Stat. Methodol.) **63**(3), 425–464 (2001)

6. Schwer, L.E.: An overview of the PTC 60/V&V 10: guide for verification and validation in computational solid mechanics. Eng. Comput. **23**(4), 245–252 (2007)
7. Roache, P.J.: Perspective: validation – What does it mean? J. Fluid. Eng. **131**(3) (2009)
8. Roache, P.J.: Fundamentals of Verification and Validation. Hermosa Publishers, Socorro (2009)
9. Dvurecenska, K., Graham, S., Patelli, E., Patterson, E.A.: A probabilistic metric for the validation of computational models. R. Soc. Open Sci. **5**(11), 180687 (2018)
10. Liu, Y., Chen, W., Arendt, P., Huang, H.Z.: Toward a better understanding of model validation metrics. J. Mech. Des. **133**(7) (2011)
11. Gardner, P., Lord, C., Barthorpe, R.J.: A unifying framework for probabilistic validation metrics. J. Verification Validation Uncertain. Quantif. **4**(3) (2019)
12. Ferson, S., Oberkampf, W.L., Ginzburg, L.: Model validation and predictive capability for the thermal challenge problem. Comput. Method. Appl. Mech. Eng. **197**(29), 2408–2430 (2008)
13. Rebba, R., Huang, S., Liu, Y., Mahadevan, S.: Statistical validation of simulation models. Int. J. Mat. Product Technol. **25**(1–3), 164–181 (2006)
14. Sindo, K.: Welding Metallurgy. John Wiley and Sons, Inc., New Jersey (2003)
15. American Welding Society. Structural Welding Code – Steel. AWS, Miami (1994)
16. Kreye, H.: Melting phenomena in solid state welding processes. Weld. J. **56**(5), 154–158 (1977)
17. Benaroya, H., Nagurka, M., Han, S.: Mechanical Vibration: Analysis, Uncertainties, and Control. CRC Press, Boca Raton (2017)
18. Bauchau, O.A., Craig, J.I.: Euler-Bernoulli beam theory. In: Structural Analysis, pp. 173–221. Springer, Berlin (2009)
19. Nashif, A.D., Jones, D.I.G., Henderson, J.P.: Vibration Damping. John Wiley & Sons, Hoboken (1985)
20. Orban, F.: Damping of materials and members in structures. In: Journal of Physics – Conference Series, vol. 268, pp. 012–022. IOP Publishing, Bristol (2011)
21. Peeters, B., De Roeck, G.: Reference-based stochastic subspace identification for output-only modal analysis. Mech. Syst. Signal Proces. **13**(6), 855–878 (1999)
22. Brincker, R., Andersen, P.: Understanding stochastic subspace identification. In: Proceedings of the 24th IMAC, St. Louis, vol. 126 (2006)
23. Ching, J., Chen, Y.: Transitional Markov Chain Monte Carlo method for Bayesian model updating, model class selection, and model averaging. J. Eng. Mech. **133**(7), 816–832 (2007)
24. Betz, W., Papaioannou, I., Straub, D.: Transitional Markov Chain Monte Carlo: observations and improvements. J. Eng. Mech. **142**(5), 04016016 (2016)
25. Ferson, S., Oberkampf, W.L.: Validation of imprecise probability models. Int. J. Reliab. Saf. **3**(1/2/3), 3–22 (2009)

Chapter 6
A Covariance-Based Approach for Recovering Phase-Consistent Loads from Random Vibration Analysis

T. Van Fossen

Abstract In the field of aerospace analysis, it is common practice to apply the root-mean-square (RMS) interface loads from a system-level random vibration analysis as static inputs to high-fidelity stress models. This process serves as an efficient means for evaluating several designs or design iterations; however, because the interface loads typically exhibit different magnitude and phase relationships at different times and frequencies, it is unrealistic to assume that all loads should be applied simultaneously at their maximum, in-phase magnitudes (often assumed to be the three-sigma or three times RMS loads). A more realistic approach for enveloping the problem space is to define a load vector (load case) for each interface degree of freedom (DOF), with a selected DOF applied at its maximum magnitude and all other DOFs applied at their phase-consistent magnitudes. These phase-consistent loads would seem easy to calculate through transient analysis by identifying the peak or valley with the desired magnitude for each DOF (e.g., three-sigma) and extracting the phase-consistent loads for the other DOFs; however, each peak or valley will produce a different load vector, since each peak or valley is merely a single data point in a statistical distribution. Thus, the preferred approach is to obtain phase-consistent loads in the frequency domain, such that the load vectors represent the expected magnitudes based on the statistical distribution of loads. This paper presents a frequency-domain, covariance-based approach for obtaining phase-consistent static loads from random vibration analysis. The load vectors obtained using this approach reasonably and realistically envelope the problem space and provide an efficient means for conducting high-fidelity random vibration stress analysis.

Keywords Covariance · Phase-consistent loads · Load vectors · Random vibration analysis · Stress analysis

Nomenclature

a_1	Length of the semi-major axis of a one-sigma SDE
a_n	Length of the semi-major axis of an n-sigma SDE
b_1	Length of the semi-minor axis of a one-sigma SDE
$b\text{In}_n$	Boolean indicating which data points are enclosed by an n-sigma SDE
b_n	Length of the semi-minor axis of an n-sigma SDE
COV	Covariance matrix
$\text{COV}_{i,*}$	The ith row of the covariance matrix
$\text{COV}_{i,i}$	Self-variance for DOF i
COV_{XY}	2×2 covariance matrix for DOF X and DOF Y
CSD	Cross-power spectral density
$\text{CSD}_{j,k}$	CSD between response j and response k
DOF	Degree of freedom
$f1$	Lowest frequency of interest
$f2$	Highest frequency of interest
FEM	Finite element model
FRF	Frequency response function
$\text{FRF}^*_{i,j}$	Complex conjugate transpose of the FRF for reference i and response j

T. Van Fossen (✉)
ATA Engineering, Inc., San Diego, CA, USA
e-mail: tyler.vanfossen@ata-e.com

M. S. Allen et al. (eds.), *Dynamic Substructures, Volume 4*, Conference Proceedings of the Society for Experimental Mechanics Series, https://doi.org/10.1007/978-3-030-75910-0_6

$FRF_{i,k}$	FRF for reference *i* and response *k*
LV_i	Load vector for DOF *i*
N	Number of data points in each time history
n	Number of standard deviations (sigma value)
P_n	Amount of data enclosed by an *n*-sigma SDE (as a fraction of *N*)
PSD	Power-spectral density
PSD_i	PSD for DOF *i*
R	Rotation matrix (transformation matrix)
RMS	Root-mean-square
S	Diagonal matrix of singular values for the real portion of the covariance matrix
SDE	Standard deviational ellipse
SVD	Singular value decomposition
TH_x	Time history for DOF *X* (column vector)
$TH_{x'}$	Projection of TH_x and TH_y onto the major axis of an SDE defined for DOFs *X* and *Y*
TH_y	Time history for DOF *Y* (column vector)
$TH_{y'}$	Projection of TH_x and TH_y onto the minor axis of an SDE defined for DOFs *X* and *Y*
U, V	Unitary matrices calculated so that $COV_{XY} = U * S * V'$
x_n	Projection of x'_n and y'_n onto the *x*-axis
x'_n	Row-vector of coordinates along the major axis of an *n*-sigma SDE
y_n	Projection of x'_n and y'_n onto the *y*-axis
y'_n	Row-vector of coordinates along the minor axis of an *n*-sigma SDE

6.1 Introduction

Traditional random vibration analysis is conducted in the frequency domain using a finite element model (FEM) with enough fidelity to accurately characterize the dynamic behavior of the system, but typically the FEM does not have enough fidelity to accurately recover component-level stresses. Therefore, it is common practice to recover interface loads from the system-level (dynamics) model and to apply those loads to high-fidelity breakout (stress) models to recover the component-level stresses. Three-sigma RMS (three times RMS) loads are typically used to envelope the problem space, where three-sigma corresponds to a 99.7% confidence level for a normal distribution. For each breakout model, the signed combinations of the relevant interface loads are applied as inputs, and the maximum (enveloped) stress is reported. Despite the simplicity of this approach, the resulting stresses are often overly conservative and unrealistic because the cross-correlation among loads is not considered, so all loads are simultaneously applied at their three-sigma RMS values. Additionally, the number of load cases that must be analyzed using the traditional approach may become prohibitive when the number of interface DOFs and/or breakout models is large and several designs or design iterations are to be analyzed.

This paper presents an alternative approach for conducting random vibration breakout analyses, where the covariance among interface loads is used to calculate load vectors containing phase-consistent loads that can be used to conduct a reduced number of static analyses that reasonably and realistically envelope the problem space. Only one load vector (load case) is needed for each interface DOF (two if both positive and negative loads are to be considered), where each load vector defines the signed magnitudes of all interface loads when a given DOF is at a prescribed value (e.g., its three-sigma RMS). The load vectors effectively define statistically likely combinations of loads that can be linearly scaled to achieve a desired confidence level (e.g., 99.7% confidence that the magnitudes of the interface loads will be less than or equal to the magnitudes defined by the load vectors). Because the peak magnitude of each interface DOF is captured by a load vector (one load vector per DOF), multiple critical stress states within the breakout model are reasonably exercised.[1] Thus, the covariance-based load vectors approach provides an efficient means for evaluating several designs or design iterations using realistic loads, thereby reducing the amount of conservatism and improving confidence in the stress results obtained from statically conducted random vibration analyses.

[1] As is true for all random vibration static breakout analyses, the static deflection shapes must match the dynamic (mode) shapes to exercise the structure in the same manner (i.e., produce the same stresses) between the two types of analyses.

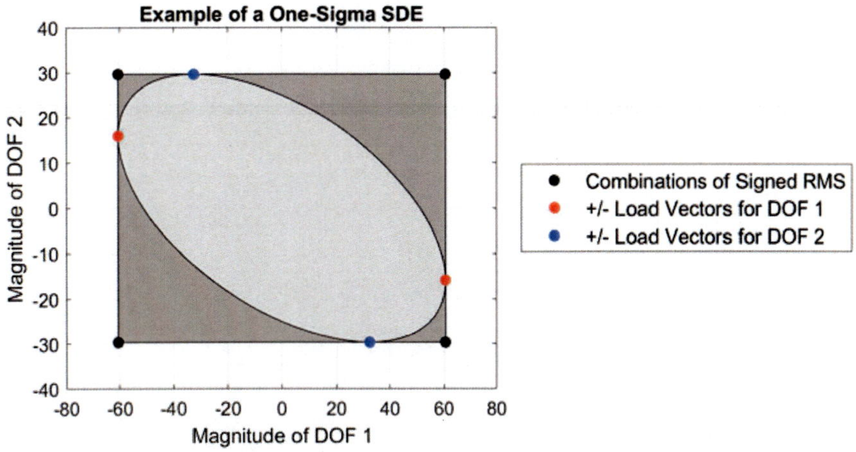

Fig. 6.1 Example of a one-sigma SDE for two negatively correlated DOFs

6.2 Background

Phase-consistent loads can be obtained from random vibration analysis using the concept of covariance to characterize the relative magnitude and phase relationships among a given set of interface loads. The symmetric covariance matrix is calculated by integrating the cross-power spectral density (CSD) matrix. The resulting matrix effectively describes how each DOF behaves relative to each other DOF. Using this information, load vectors can be calculated, wherein each load vector describes the expected, signed magnitudes of all DOFs when a particular DOF is at a prescribed value. The covariance matrix can also be used to calculate an ellipse (two DOFs), ellipsoid (three DOFs), or hyper-ellipsoid (more than three DOFs) that can be linearly scaled to bound the problem space at a given confidence level. The magnification factor of the ellipse, ellipsoid, or hyper-ellipsoid is referred to as its sigma value, where n-sigma corresponds to a magnification factor of n. Figure 6.1 shows a one-sigma standard deviational ellipse (SDE) as an example.

In this case, the ellipse is drawn through the one-sigma RMS values for both DOFs, whereas the red dots correspond to the plus-and-minus one-sigma RMS values for DOF 1 and the blue dots correspond to the plus-and-minus one-sigma RMS values for DOF 2. The orientation of the ellipse tells us that the two DOFs are negatively correlated, meaning that as the value (load) of one DOF increases, the value (load) of the other DOF generally decreases. The positive, one-sigma load vector for DOF 1 in this case is approximately $[+61, -16]$, where the first value corresponds to the one-sigma RMS of DOF 1, and the second value corresponds to the phase-consistent value (load) for DOF 2. The positive, one-sigma load vector for DOF 2 is approximately $[-32, +30]$, where the second value corresponds to the one-sigma RMS of DOF 2 and the first value corresponds to the phase-consistent value (load) for DOF 1. Each point on (and within) the ellipse can be thought of as a realistic loading situation. The signed combinations of the RMS values define the corners of a rectangular box (black dots) that encloses the ellipse—the black dots represent the load cases that are analyzed using the traditional RMS approach. Given that the corners of the box do not fall on the ellipse (two of the corners are nowhere near the ellipse), this example definitively shows that the RMS approach produces unrealistic, overly conservative load cases.

It is important to note that the amount of data (number of load combinations) enclosed by a given ellipse, ellipsoid, or hyper-ellipsoid (i.e., its confidence level) does not follow the traditional 68–95–99.7 rule, since the empirical rule applies only to univariate distributions. For multivariate distributions, the confidence level at a given magnification factor (analogous to sigma value) is calculated from the chi-squared distribution for the applicable number of DOFs [1], such that the confidence level for a given sigma value decreases as the number of DOFs increases. Although this seems to imply that very high sigma values may be needed to envelope the problem space for interfaces containing many DOFs, the number of DOFs driving each stress state enveloped by the load vectors is typically very low, so three-sigma load vectors are sufficient for most applications.

Table 6.1 shows the confidence levels as a function of sigma value and the number of active DOFs (i.e., the number of DOFs driving a critical stress state). Note that the confidence levels for the one-DOF case follow the empirical rule, since this case represents a univariate distribution.

Time history data (obtained from CSDs, transient analysis, or test measurements) are not needed to calculate the load vectors but help us to understand the physical meaning of the ellipses. For the two-dimensional case, each time history (one time history per DOF) can be plotted relative to each other time history and compared to the applicable SDEs. The amount of data enclosed by each SDE is expected to match the confidence levels associated with a bivariate distribution. Figure 6.2 shows an example with six incremental SDEs (one sigma through six sigma) with equivalent time history data generated

Table 6.1 Confidence levels as a function of the number of DOFs and the scaling values

	Magnification factor (analogous to sigma value)						
# Active DOFs	1	2	3	4	5	6	7
1	0.6827	0.9545	0.9973	0.9999	1.0000	1.0000	1.0000
2	0.3935	0.8647	0.9889	0.9997	1.0000	1.0000	1.0000
3	0.1987	0.7385	0.9707	0.9989	1.0000	1.0000	1.0000
4	0.0902	0.5940	0.9389	0.9970	0.9999	1.0000	1.0000
5	0.0374	0.4506	0.8909	0.9932	0.9999	1.0000	1.0000
6	0.0144	0.3233	0.8264	0.9862	0.9997	1.0000	1.0000

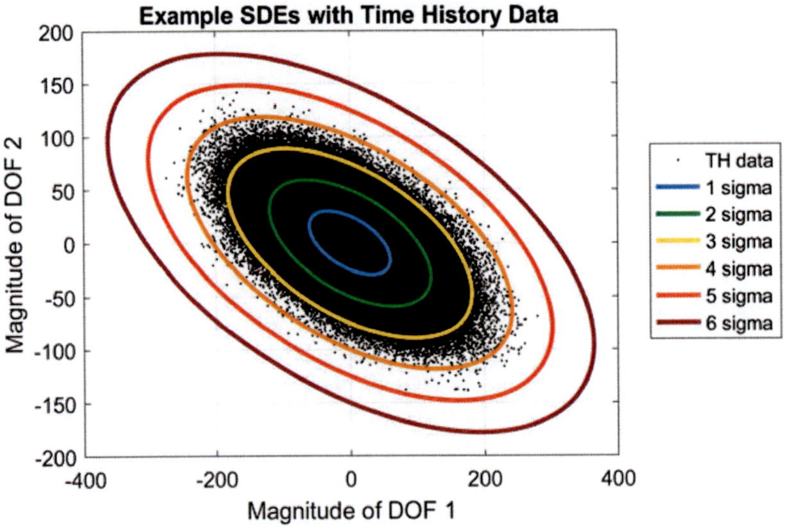

Fig. 6.2 Example SDEs with overlaid time history data (approximately 275,000 peaks and valleys)

Table 6.2 Amount of data enclosed by each SDE in Fig. 6.2

	Amount of data enclosed (confidence level)		
# Standard deviations	Expected (bivariate)	Actual	Difference
1	0.3935	0.4082	0.0147
2	0.8647	0.8856	0.0209
3	0.9889	0.9930	0.0041
4	0.9997	1.0000	0.0003
5	1.0000	1.0000	0.0000

from the same CSDs used to calculate the SDEs. Note that only the union of the peaks and valleys for the two time histories is plotted to help visualize data density. For this example, the time histories contain approximately ten million samples (time points) but fewer than 275,000 peaks and valleys. Using this plotting technique, it is evident that the density of the peaks and valleys decreases rapidly beyond the three-sigma SDE, as expected.

Table 6.2 presents the amount of time history data (all data—not just peaks and valleys) enclosed by each ellipse in Fig. 6.2 and compares the results to the expected confidence levels for a bivariate distribution. The amount of data enclosed by each SDE is within approximately 2% of its expected amount, suggesting that the finite-duration time histories capture the expected distribution of loads reasonably well. Small differences are expected, since the time history data is merely a snapshot of the expected distribution.

6.3 Technical Approach

The covariance matrix, which characterizes the expected magnitude and phase relationships among a set of responses with known cross-correlation, is calculated by integrating the relevant CSD matrix. Equation 6.1 shows the calculation of the

CSD matrix, which is essentially the product of each pair of frequency response functions (FRFs). In this case, the FRFs are defined on a per-unit-of-input basis, so the FRF product is also multiplied by the power-spectral density (PSD) of the reference DOF used to calculate the FRFs to achieve proper scaling.

Response CSD matrix:

$$\mathrm{CSD}_{j,k} = \mathrm{FRF}^{*}_{i,j} \cdot \mathrm{FRF}_{i,k} \cdot \mathrm{PSD}_{i} \tag{6.1}$$

where

$\mathrm{CSD}_{j,\,k}$ is the CSD between response j and response k
$\mathrm{FRF}^{*}_{i,j}$ is the complex conjugate transpose of the FRF for reference i and response j
$\mathrm{FRF}_{i,\,k}$ is the FRF for reference i and response k
PSD_{i} is the PSD for DOF i

The covariance matrix is then calculated using Eq. 6.2, where the CSD matrix is integrated over the frequency range of interest. For the examples presented in this paper, this integration was performed using the *cumtrapz* function in MATLAB, retaining only the final integration values (cumulative sums) in the covariance matrix. Note that the *cumtrapz* function assumes linear interpolation (lin-lin), but logarithmic interpolation (log-log) can be approximated by using high-resolution CSDs that were calculated using log-log interpolation. The interpolation method used to integrate the CSD matrix must match the method used to calculate the RMS values for the load vectors to be defined at the intended RMS values.

Covariance matrix:

$$\mathrm{COV} = \int_{f1}^{f2} \mathrm{CSD} \tag{6.2}$$

where

CSD is the response CSD matrix
COV is the covariance matrix
$f1$ is the lowest frequency of interest
$f2$ is the highest frequency of interest

Load vectors are calculated from the covariance matrix by dividing each row or column by the square root of the applicable diagonal term. While the directionality of the load vectors calculation does not affect the results (since the covariance matrix is symmetric), the same direction must be used to correctly extract the load vectors from the resulting matrix. In other words, if the load vectors are calculated on a row-by-row basis, each row of the resulting load vectors matrix defines a load vector. On the other hand, if the load vectors are calculated on a column-by-column basis, each column of the resulting load vectors matrix defines a load vector. Equation 6.3 shows the load vectors calculation using the row-by-row approach. The square-root operator is necessary because the covariance matrix is calculated from cross-power quantities (squared magnitudes).

Load vectors (row by row):

$$\mathrm{LV}_{i} = \mathrm{real}\left(\mathrm{COV}_{i,*}\right) / \sqrt{\mathrm{COV}_{i,i}} \tag{6.3}$$

where

$\mathrm{COV}_{i,\,i}$ is the self-variance for DOF i
$\mathrm{COV}_{i,\,*}$ is the ith row of the covariance matrix
LV_{i} is the load vector for DOF i

Each load vector effectively defines the expected magnitude and phase (sign) for each DOF when a given DOF is at a prescribed value. Prior to scaling, each diagonal term is equal to the one-sigma RMS for the applicable DOF (one row or column per DOF), and the off-diagonal terms represent the corresponding phase-consistent values. The absolute values of the off-diagonal terms for a given DOF are always less than or equal to the applicable diagonal term. The load vectors can be linearly scaled to obtain phase-consistent loads at a given sigma value. In other words, three-sigma load vectors (diagonal terms equal to three times RMS) can be obtained by multiplying each load vector by a factor of three. An example set of three-sigma load vectors is show in Fig. 6.3.

For this example, there are six interface DOFs and, thus, six corresponding load vectors (12 after considering both positive and negative load vectors). Each load vector is applied to the breakout model as a separate static load case, and the resulting

Phase-Consistent Loads for Each Load Vector

Three-Sigma Load Vectors	DOF 1	DOF 2	DOF 3	DOF 4	DOF 5	DOF 6
DOF 1	181.8	-47.7	16.5	-400.5	-12.7	-14.1
DOF 2	-97.5	89.0	-28.1	421.5	72.1	-112.7
DOF 3	37.1	-30.9	80.8	-93.2	-2.3	-3.0
DOF 4	-60.4	31.1	-6.2	1206.4	153.0	-176.8
DOF 5	-10.1	27.9	-0.8	803.2	229.7	-213.7
DOF 6	-9.6	-37.7	-0.9	-802.6	-184.7	265.8

← Phase-consistent loads when DOF 2 = 3*RMS

Max magnitude is on the diagonal (true for each DOF)

Diagonal = 3*RMS

Fig. 6.3 Example three-sigma load vectors for a set of six DOFs

stresses are enveloped to identify the stresses used to calculate the margins of safety. Because one load vector is defined for each interface DOF with that DOF at its maximum expected value, the load vectors reasonably exercise all critical stress states for the component (assuming the static deflection shapes are a good approximation of the mode shapes driving each stress state; see footnote 1). Using the traditional approach for this same example, 64 (2^6) load cases would be needed to analyze all of the signed combinations of three-sigma interface loads, and because the interface loads are combined in an unrealistic manner, the resulting (enveloped) stresses would be much more conservative than the stresses obtained from the load vectors approach.

6.4 Visualizing the Problem Space

While load vectors can be calculated without plotting any ellipses, ellipsoids, or hyper-ellipsoids, these plots are invaluable for visualizing the problem space and validating the results. Thus, for completeness, this section presents the methodology for plotting the results, focusing on the two-dimensional case for simplicity, although the methodology is essentially the same for the multi-dimensional case after adjusting the equations to account for the extra dimension(s). Ellipsoids and hyper-ellipsoids can be visualized in two dimensions by plotting the ellipses calculated for each pair of DOFs, where each ellipse corresponds to the projection of the resulting surface onto each plane.

Figure 6.4 shows a three-dimensional example to demonstrate this concept. The three ellipses (black lines) do not define points on the surface of the ellipsoid, but rather the projection of the ellipsoid onto each plane. The load vectors define the points on the surface (black dots) corresponding to the maximum and minimum values for each DOF. The bounding box defined by the signed combinations of the extrema (not shown) completely encloses the ellipsoid, intersecting only at the points used to define the load vectors. The ellipsoid is arbitrarily colored/shaded based on the magnitude of DOF 3 for the sole purpose of making it easier to visualize the surface in three dimensions.

Equation 6.4 shows that a one-sigma SDE can be completely defined (magnification and orientation) by the singular value decomposition (SVD) of the real portion of the applicable covariance matrix. The singular values can be thought of as the squared magnitudes of the semi-major and semi-minor axes—squared magnitudes because the covariance matrix contains power quantities. Thus, the square root of the singular values defines the magnitudes of the semi-major and semi-minor axes. Since the covariance matrix is always symmetric, the two unitary matrices are identical, so both matrices define the orientation of the principal axes (i.e., the rotation matrix). The rotation matrix is used to convert between the global (X-Y) and local (ellipse) coordinate systems in Eqs. 6.5 and 6.6, which are detailed later in this section.

Size and orientation of a one-sigma SDE:

$$[U, S, V] = svd\,(\text{real}\,(\text{COV}_{XY})) \tag{6.4a}$$

$$a_1 = \sqrt{S_{1,1}} \tag{6.4b}$$

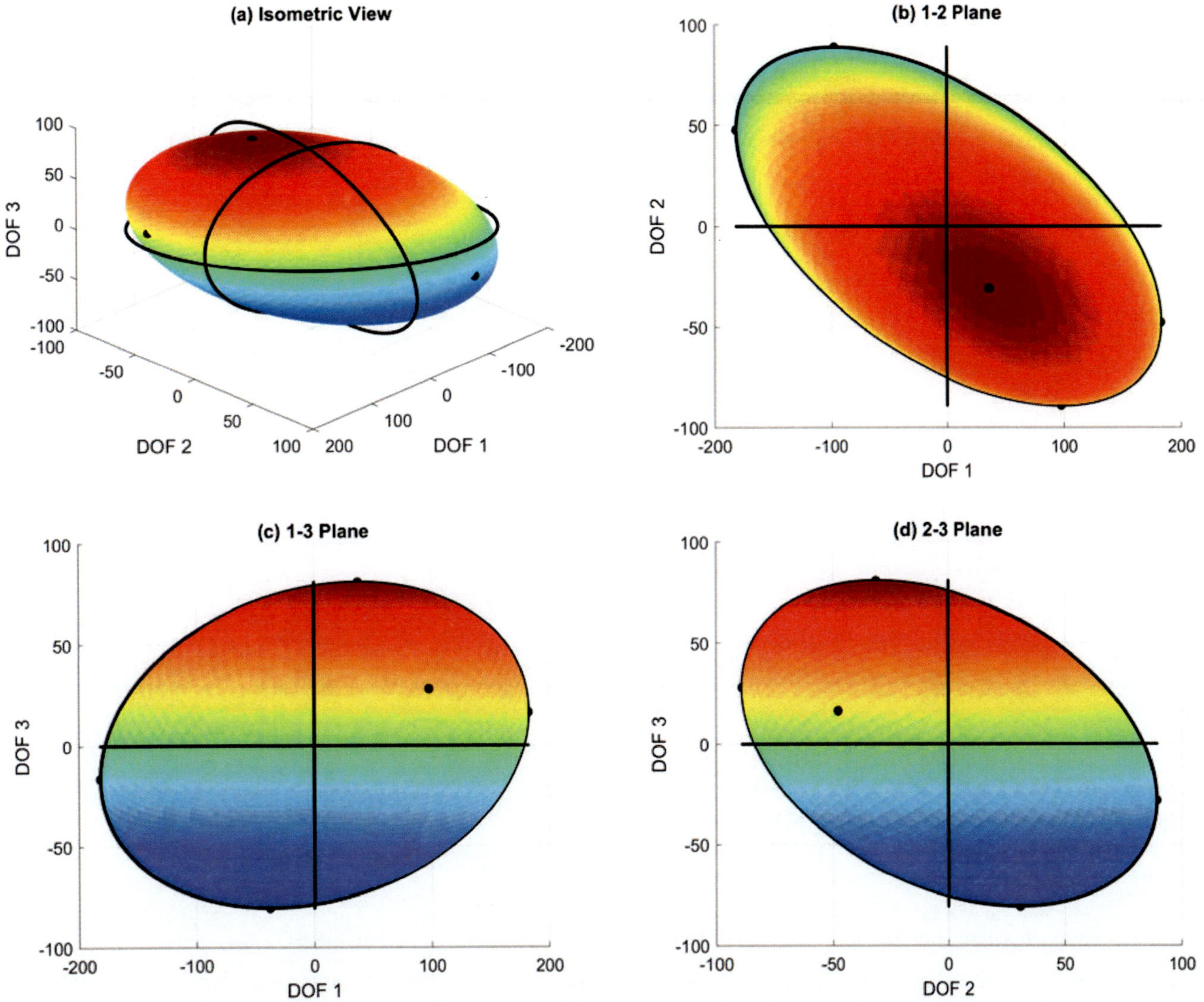

Fig. 6.4 Example of a three-sigma standard deviational ellipsoid with corresponding two-dimensional ellipses (black lines) and load vectors (black dots): (**a**) isometric view, (**b**) projection onto the 1-2 plane, (**c**) projection onto the 1-3 plane, and (**d**) projection onto the 2-3 plane

$$b_1 = \sqrt{S_{2,2}} \tag{6.4c}$$

$$R = U = V \tag{6.4d}$$

where

a_1 is the length of the semi-major axis of a one-sigma SDE
b_1 is the length of the semi-minor axis of a one-sigma SDE
COV_{XY} is the 2×2 covariance matrix for DOF X and DOF Y
R is the rotation matrix (transformation matrix)
S is the diagonal matrix of singular values for the real portion of the covariance matrix
U, V are unitary matrices calculated so that $\mathrm{COV}_{XY} = U * S * V'$

A one-sigma SDE can be scaled to an n-sigma SDE by and semi-minor axes by the desired number of standard deviations, n. The orientation of the ellipse does not change, but the maximum values on the ellipse scale proportionally with n. In other words, the maximum values on an n-sigma SDE always correspond to the n times RMS values for the DOFs used to calculate the ellipse. Equation 6.5 shows the method for calculating the points on an n-sigma SDE. These points can be transformed into the X-Y coordinate system by pre-multiplying the local coordinates (i.e., the coordinate system of the ellipse) by the applicable rotation matrix.

Points along an n-sigma SDE:

$$a_n = n * a_1 \tag{6.5a}$$

$$b_n = n * b_1 \tag{6.5b}$$

$$y'_n = \pm\sqrt{{b_n}^2 * (1 - {x'_n}^2/{a_n}^2)} \tag{6.5c}$$

$$\begin{bmatrix} x_n \\ y_n \end{bmatrix} = R * \begin{bmatrix} x'_n \\ y'_n \end{bmatrix} \tag{6.5d}$$

where

a_n is the length of the semi-major axis of an n-sigma SDE
b_n is the length of the semi-minor axis of an n-sigma SDE
n is the number of standard deviations (sigma value)
x_n is the projection of x'_n and y'_n onto the x-axis
x'_n is a row-vector of coordinates along the major axis of an n-sigma SDE
y_n is the projection of x'_n and y'_n onto the y-axis
y'_n is a row-vector of coordinates along the minor axis of an n-sigma SDE

Figure 6.5 shows the incremental ellipses (one sigma through six sigma) for the same dataset used to generate the three-sigma load vectors presented in Fig. 6.3, along with relevant time history data for demonstration purposes. Each subplot represents one element of the first load vector (i.e., the load vector for DOF 1), where the phase-consistent loads for DOFs 2–6 are identified from the three-sigma ellipses by finding the point on each ellipse corresponding the three-sigma value for DOF 1. For example, when DOF 1 is at its three-sigma RMS value, the values of the other DOFs are at −47.7 (DOF 2), 16.5 (DOF 3), −400.5 (DOF 4), −12.7 (DOF 5), and −14.1 (DOF 6). The upper left subplot, as expected, shows that DOF 1 behaves linearly with itself, while each of the other subplots shows the covariance between DOF 1 and each of the other DOFs. Although it is difficult to visualize bodies in more than three dimensions, each ellipse represents the projection of a six-dimensional hyper-ellipsoid onto each relevant two-dimensional plane.

Using Eq. 6.6, the amount of time history data enclosed by an n-sigma SDE is calculated by transforming the data to the local coordinate system of the ellipse, plugging the transformed data into the equation for the ellipse, and comparing the resulting values to unity. A data point is within the ellipse if its resultant is less than one, on the ellipse if its resultant is equal to one, and outside the ellipse if its resultant is greater than one. Therefore, the amount of data enclosed by the ellipse (including the points that are on the ellipse) is calculated by counting the number of data points with resultants less than or equal to one. The percentage of data enclosed by the ellipse is then the number of data points enclosed by the ellipse divided by the total number of data points. Since an ellipse is two-dimensional, the percentages are expected to match the confidence levels for a bivariate distribution (see Table 6.1 for two DOFs), assuming the time histories are of sufficient length to properly capture the distribution of loads.

Amount of time history data enclosed by an n-sigma SDE:

$$\begin{bmatrix} \mathrm{TH}_{x'} \\ \mathrm{TH}_{y'} \end{bmatrix} = \begin{bmatrix} \mathrm{TH}_{x} \\ \mathrm{TH}_{y} \end{bmatrix} * R \tag{6.6a}$$

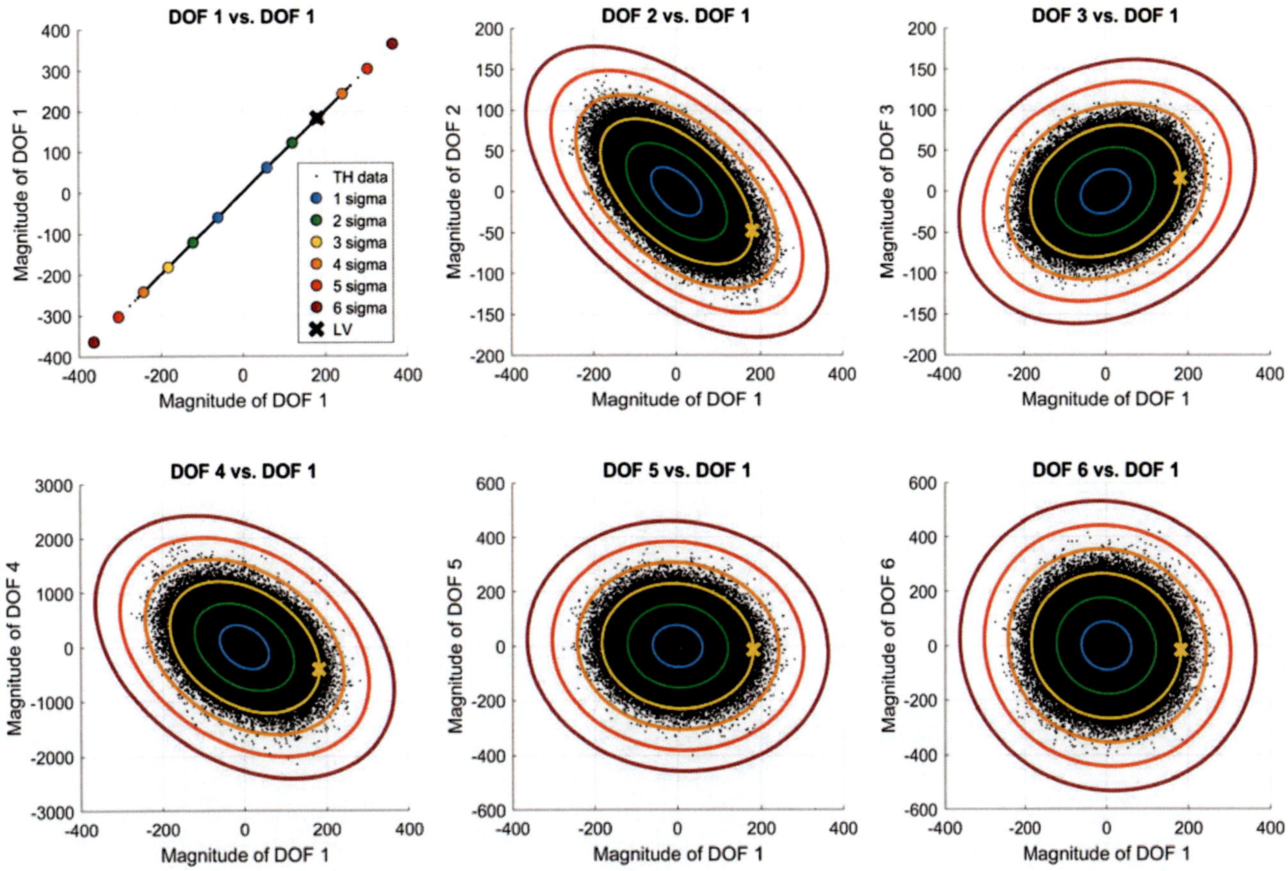

Fig. 6.5 Example load vector for DOF 1 with relevant SDEs and time history data

$$b\mathrm{In}_n = \mathrm{TH}_{x'}{}^2/{a_n}^2 + \mathrm{TH}_{y'}{}^2/{b_n}^2 \leq 1 \tag{6.6b}$$

$$P_n = \mathrm{sum}\,(b\mathrm{Inside})\,/n \tag{6.6c}$$

where

$b\mathrm{In}_n$ is a Boolean indicating which data points are enclosed by an n-sigma SDE
N is the number of data points in each time history
P_n is the amount of data enclosed by an n-sigma SDE (as a fraction of N)
$\mathrm{TH_x}$ is the time history for DOF X (column vector)
$\mathrm{TH}_{x'}$ is the projection of $\mathrm{TH_x}$ and $\mathrm{TH_y}$ onto the major axis of an SDE defined for DOFs X and Y
$\mathrm{TH_y}$ is the time history for DOF Y (column vector)
$\mathrm{TH}_{y'}$ is the projection of $\mathrm{TH_x}$ and $\mathrm{TH_y}$ onto the minor axis of an SDE defined for DOFs X and Y

6.5 Conclusion

By accounting for the covariance among interface loads, a reduced number of static breakout analyses may be conducted to obtain realistic component-level stresses from a system-level random vibration analysis. Each component-level static load case, referred to as a load vector, defines the statistically likely, signed magnitudes of the interface loads when a particular

DOF is at a prescribed value (e.g., its three-sigma RMS value). The signed magnitudes are calculated from the covariance matrix, such that the general phasing trends among the interface loads are accounted for. Just one load vector is required per interface DOF (or two if both positive and negative load vectors are to be considered), with each load vector capturing the maximum (statistically likely) magnitude for a given DOF. Each load vector may exercise a different stress state using a realistic combination of loads, and the load vectors collectively (and realistically) envelope the problem space. Compared to the traditional approach, where each signed combination of three-sigma interface loads is applied as a separate load case, the covariance-based load vectors approach significantly reduces the number of load cases required, particularly when the number of interface DOFs is large.

The same covariance matrix used to calculate the load vectors can also be used to visualize the problem space, where the SVD of the covariance matrix defines the magnification and orientation of an ellipse (two dimensions), ellipsoid (three dimensions), or hyper-ellipsoid (more than three dimensions) that encloses a certain amount of data (number of load combinations) at a given confidence level. Each point on and within the elliptical shape represents a realistic combination of interface loads, with each load vector defining the point on the surface where a given DOF is at its maximum magnitude. Since the points defined by the signed combinations of RMS interface loads are not located on the surface of the elliptical shape, it is clear that the traditional approach defines unrealistic load cases. Thus, compared to the traditional RMS-based approach, the covariance-based load vectors approach produces more realistic stresses using a reduced number of load cases, thereby improving the accuracy and efficiency of statically conducted component-level random vibration stress analyses.

Acknowledgments A special thank you to Dr. Kenneth Smith for developing the presented load vectors approach, sharing the approach with us, and allowing us to publish it. Given that the concept of covariance is not new, it is likely that similar implementations exist; however, the author is unaware of any implementations specific to obtaining phase-consistent loads from random vibration analysis.

Reference

1. Wang, B., Shi, W., Miao, Z.: Confidence analysis of standard deviational ellipse and its extension into higher dimensional Euclidean space. PLoS One. **10**(3), e0118537 (2015)

Chapter 7
On the Estimation of Structural Admittances from Near-Field Acoustic Holography Using a Dynamic Substructuring Approach

Gregor Čepon, Domen Ocepek, Miha Kodrič, and Miha Boltežar

Abstract Near-field acoustic holography (NAH) provides a reconstruction of the normal particle velocity radiated by the source of interest on a nearby surface. Reconstructing structural admittance based on normal velocity establishes NAH as alternative to the conventional methods such as accelerometers or laser vibrometers. Measuring vibrating object using microphone array offers high-resolution measurements and with this the possibility to obtain spatially dense displacement field of the structure. Although useful, the method's ill posed nature needs to be considered which introduces numerical noise to the reconstructed result. In this paper an alternative approach for improved estimation of FRFs from NAH is introduced by using dynamic substructuring methodology. Recently developed System Equivalent Model Mixing (SEMM) method is proposed that enables mixing of response models of the same structure into one hybrid model. In the hybrid formulation, each model contributes its own advantages to yield best combination of the two. NAH introduces a high spatial resolution to the hybrid model whereas discrete laser vibrometer measurements improve the accuracy of the reconstructed displacement field. Experimental validation of methodology is performed on a freely supported structure that is excited using electrodynamic shaker. Noisy acoustical environment introduces spurious peaks in FRFs acquired using NAH. The laser vibrometer measurement is not affected by a noisy environment, as only the structural response is measured. It is shown that hybrid model provides a more consistent FRFs of the structural response, especially with regard to the amplitude in the resonance regions.

Keywords Near-field acoustic holography · Noisy environment · Frequency-based substructuring · System equivalent model mixing · Hybrid model

7.1 Introduction

Near-field acoustic holography (NAH) [1] is a powerful tool for the reconstruction of acoustic quantities such as acoustic pressure, acoustic intensity, acoustic power, and normal surface velocity based on an acoustic pressure measurement called a hologram. This non-contact method is based on the source information that is captured in the near field of an acoustic source. A fine grid of microphone positions enables a spatially dense measurement and with this the possibility to obtain a high-resolution structural response.

The overall accuracy of the NAH method is strongly influenced by the measurement environment and the presence of background noise. Therefore an alternative approach is presented to improving the consistency of the structural admittances obtained from the planar NAH. In contrast to other regularization and/or field-separation techniques, which rely on the measurement of acoustic quantities, here the estimation of structural admittances is carried out by including measurement of the response on the structure itself. The methodology proposes the integration of the response model of the NAH measurement with accurate discrete measurements (e.g., accelerometer or laser vibrometer) using dynamic substructuring (DS) techniques. This is established using the recently developed System Equivalent Model Mixing (SEMM) method [2]. With SEMM, different dynamic models of the same system can be mixed into a single hybrid model. Hybrid approach offers a more consistent measurement of the structural response, especially with regard to the amplitude in the resonance regions and spurious dynamics introduced by a noisy environment.

G. Čepon (✉) · D. Ocepek · M. Kodrič · M. Boltežar
Faculty of Mechanical Engineering, University of Ljubljana, Ljubljana, Slovenia
e-mail: gregor.cepon@fs.uni-lj.si; domen.ocepek@fs.uni-lj.si; miiha.kodric@fs.uni-lj.si; miha.boltezar@fs.uni-lj.si

M. S. Allen et al. (eds.), *Dynamic Substructures, Volume 4*, Conference Proceedings of the Society for Experimental Mechanics Series, https://doi.org/10.1007/978-3-030-75910-0_7

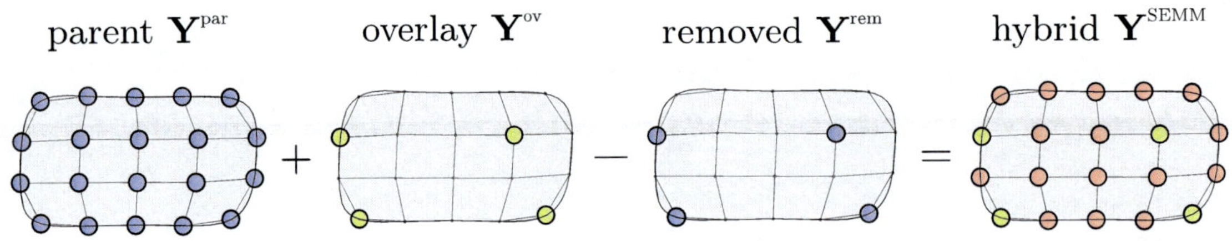

Fig. 7.1 Schematic representation of System Equivalent Model Mixing

7.1.1 System Equivalent Model Mixing

With SEMM, different equivalent dynamic models either of numerical or experimental nature can be mixed into a single hybrid model based on the Lagrange-multiplier frequency-based substructuring (LM FBS) method [3]. A schematic representation of SEMM is depicted in Fig. 7.1. First, coupling of parent and overlay model projects dynamic properties of the latter on the full-span DoF of the parent model. To remove dynamics of the parent from the newly formed model, removed model (a condensed version of parent) is decoupled. These two steps can be written as follows:

$$\mathbf{u} = \mathbf{Y}(\mathbf{f} + \mathbf{g}); \quad \text{where} \quad \mathbf{Y} = \begin{bmatrix} \mathbf{Y}^{\text{par}} & & \\ & -\mathbf{Y}^{\text{rem}} & \\ & & \mathbf{Y}^{\text{ov}} \end{bmatrix}, \quad \mathbf{f} = \begin{bmatrix} \mathbf{f}^{\text{par}} \\ \mathbf{f}^{\text{rem}} \\ \mathbf{f}^{\text{ov}} \end{bmatrix}, \quad \mathbf{g} = \begin{bmatrix} \mathbf{g}^{\text{par}} \\ \mathbf{g}^{\text{rem}} \\ \mathbf{g}^{\text{ov}} \end{bmatrix}. \tag{7.1}$$

Imposing compatibility and equilibrium conditions in accordance with the LM FBS, which ensure that the coupled structures have the same displacements and interface forces at the collocated boundary DoFs of the parent and the overlay model, following formulation is obtained:

$$\mathbf{Y}^{\text{SEMM}} = \mathbf{Y}^{\text{par}} - \mathbf{Y}^{\text{par}} \left(\begin{bmatrix} \mathbf{Y}_{\text{bi}} & \mathbf{Y}_{\text{bb}} \end{bmatrix}^{\text{rem}} \right)^{+} \left(\mathbf{Y}_{\text{bb}}^{\text{rem}} - \mathbf{Y}^{\text{ov}} \right) \left(\begin{bmatrix} \mathbf{Y}_{\text{ib}} \\ \mathbf{Y}_{\text{bb}} \end{bmatrix}^{\text{rem}} \right)^{+} \mathbf{Y}^{\text{par}}, \tag{7.2}$$

where b is regarded as the boundary and i as the internal DoF (property of the parent model only).

7.2 Estimation of Structural Admittances from NAH Using SEMM

To increase the consistency of structural admittances from NAH, a combination of different equivalent models is proposed. A schematic representation of the procedure is depicted in Fig. 7.2. First the structural velocity is validated on the acoustic source plane using NAH methodology [1]. Care should be taken to properly regularize the solution [4]. Structural admittance is then calculated based on reconstructed velocity field and excitation force. These FRFs from NAH are mixed with the FRFs from laser vibrometer using SEMM [2] in order to project consistent dynamic properties from the laser measurement on the reconstructed field from NAH.

7.3 Experiment

The proposed methodology was applied on a T-shaped structure shown in Fig. 7.3, excited using electrodynamic shaker LDS V101. The parent model $\mathbf{Y}^{\text{par}}$ was acquired using array of PCB 378B02 microphones. Additional speaker was added to the experimental setup, introducing spurious dynamic to the parent model. In total, 63 responses along top surface of the structure were measured and projected on the source plane using NAH formulation [1]. Responses measured using laser vibrometer in 3 DoF were used as the overlay model $\mathbf{Y}^{\text{ov}}$. One laser measurement was treated as a reference. Both identified response models were mixed into a hybrid model using the fully extended SEMM method (Eq. 7.2).

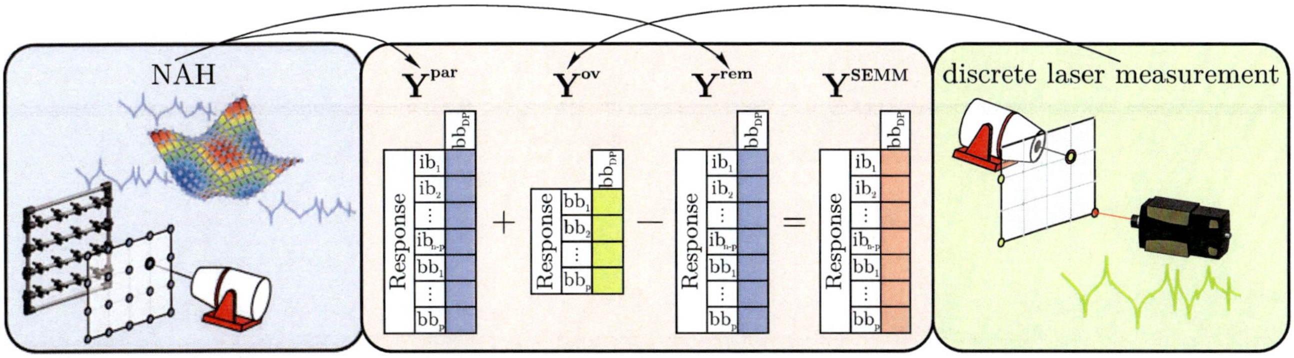

Fig. 7.2 Schematic representation of FRFs estimation using SEMM

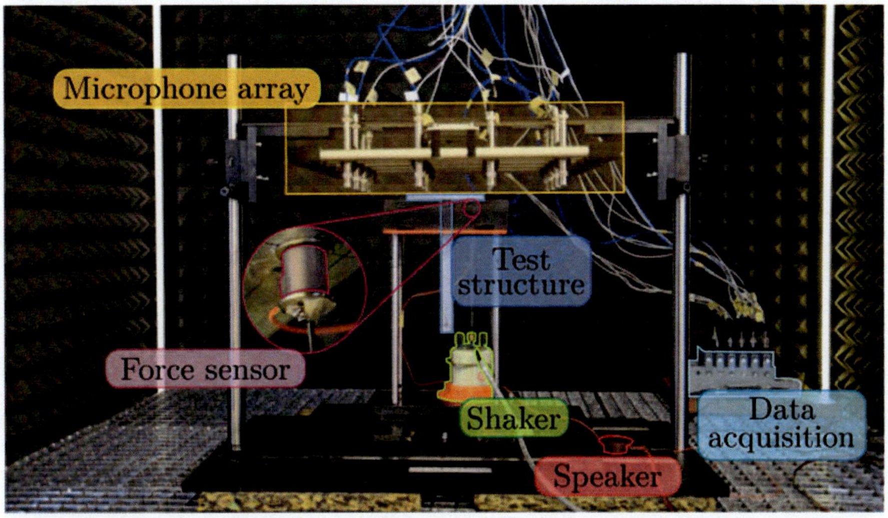

Fig. 7.3 Experimental setup of the microphone array

7.4 Results

Magnitude of the estimated hybrid FRFs is depicted in Fig. 7.4. Reconstructed FRFs from NAH and reference measurement from laser vibrometer are added for comparison. It can be observed that NAH-based measurement is affected by noisy environment. However, when using the proposed approach, this spurious acoustic source is effectively canceled out from the response of the structure. Removal of this spurious peak originates from the definition of the removed model. The latter, originating from the parent model, includes this spurious dynamics. As the removed model is decoupled from the model resulting from coupling parent and overlay models, any dynamic behavior that is not involved in the overlay model is also decoupled from the hybrid model. Also, more accurate prediction of the hybrid formulation can be clearly observed in resonant areas.

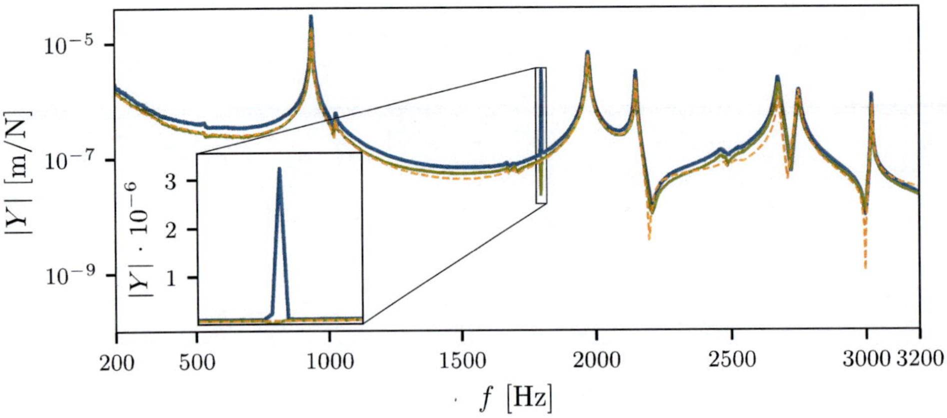

Fig. 7.4 FRFs comparison at the reference point. (—·—) NAH, (——) SEMM-NAH, (- - -) Reference

References

1. Maynard, J.D., Williams, E.G., Lee, Y.: Nearfield acoustic holography: I. Theory of generalized holography and the development of NAH. J. Acoust. Soc. Am. **78**(4), 1395–1413 (1985)
2. Klaassen, S.W., van der Seijs, M.V., de Klerk, D.: System equivalent model mixing. Mech. Syst. Signal Proces. **105**, 90–112 (2018)
3. de Klerk, D., Rixen, D.J., Voormeeren, S.: General framework for dynamic substructuring: history, review and classification of techniques. AIAA J. **46**(5), 1169–1181 (2008)
4. Williams, E.G.: Regularization methods for near-field acoustical holography. J. Acoust. Soc. Am. **110**(4), 1976–1988 (2001)

Chapter 8
Experimental Framework for Identifying Inconsistent Measurements in Frequency-Based Substructuring

Miha Kodrič, Gregor Čepon, and Miha Boltežar

Abstract The dynamic properties of modern products are analyzed using an experimental approach through the measurement of frequency-response functions (FRFs). For an individual measurement, the coherence offers an online check during the system acquisition. More general tools for determining the consistency of the complete measurement set are based on a comparison of the FRFs or the modal shapes with a numerical model. They are useful tools, but they rely on a comparison with a numerical model that might not reflect the behavior of the actual system. This paper aims to develop a comprehensive experimental method to check the consistency of individual measurements based on comparisons with the complete experimental response model. The numerical model is introduced only to enable the experimental model to be expanded using the System Equivalent Model Mixing method. The entire formulation is developed in the frequency domain, so that the transition to the modal domain, which might remove the physically relevant information from the system, is not required. In the frequency domain, it is possible to assess the consistency of the FRF across the entire frequency range of interest and not only in the region of the natural frequencies. This is of great importance in the area of frequency-based substructuring, where even small inaccuracies in the substructure's FRFs (e.g., the position of the anti-resonance) can lead to erroneous coupling results due to the inversion process. The experimental case study demonstrates the efficiency of the proposed approach. By removing the identified inconsistent measurements, it was possible to significantly increase the accuracy of the final coupling process.

Keywords Data consistency · System equivalent model mixing · Dynamic substructuring · Frequency domain · Expansion process

8.1 Introduction

Providing a consistent experimental response model is a crucial step for accurate dynamic analysis. The quality of the experimental model can be verified using comparative criteria like FRAC, MAC, CoMAC, etc. They are not able to identify the relationship between the measurements inside the experimental data set. These tools only allow a comparison with other data sets, usually obtained by a numerical model that might not reflect the behavior of the actual system. Recently, a new, general approach called the Data Consistency Assessment Function (DCAF) [1] was proposed that is able to evaluate the consistency of a set of measurements. The method is formulated in the modal domain using the SEREP expansion process and the MAC criterion. This paper presents a new approach to the identification of inconsistent measurements. The basic idea is similar to the DCAF method; however, here the entire formulation is developed within the frequency domain using System Equivalent Model Mixing (SEMM) method [2] and coherence criteria. To present the capability of the introduced method, a coupling of two simple beam-like structures is presented. It will be shown that by removing the inconsistent FRFs from the experimental substructure model, it is possible to improve the accuracy of the coupling process.

M. Kodrič (✉) · G. Čepon · M. Boltežar
Faculty of Mechanical Engineering, University of Ljubljana, Ljubljana, Slovenia
e-mail: miha.kodric@fs.uni-lj.si; gregor.cepon@fs.uni-lj.si; miha.boltezar@fs.uni-lj.si

M. S. Allen et al. (eds.), *Dynamic Substructures, Volume 4*, Conference Proceedings of the Society for Experimental Mechanics Series, https://doi.org/10.1007/978-3-030-75910-0_8

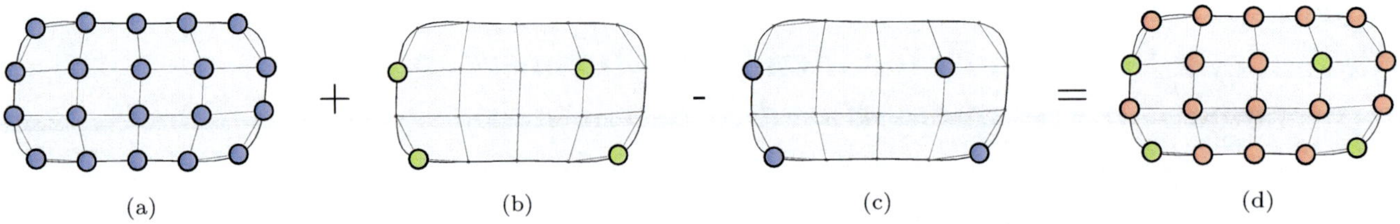

Fig. 8.1 Equivalent models for SEMM method; (**a**) Parent model $\mathbf{Y}^{\text{par}}$, (**b**) Overlay model $\mathbf{Y}^{\text{ov}}$, (**c**) Removed model $\mathbf{Y}^{\text{rem}}$, (**d**) Hybrid model $\mathbf{Y}^{\text{SEMM}}$

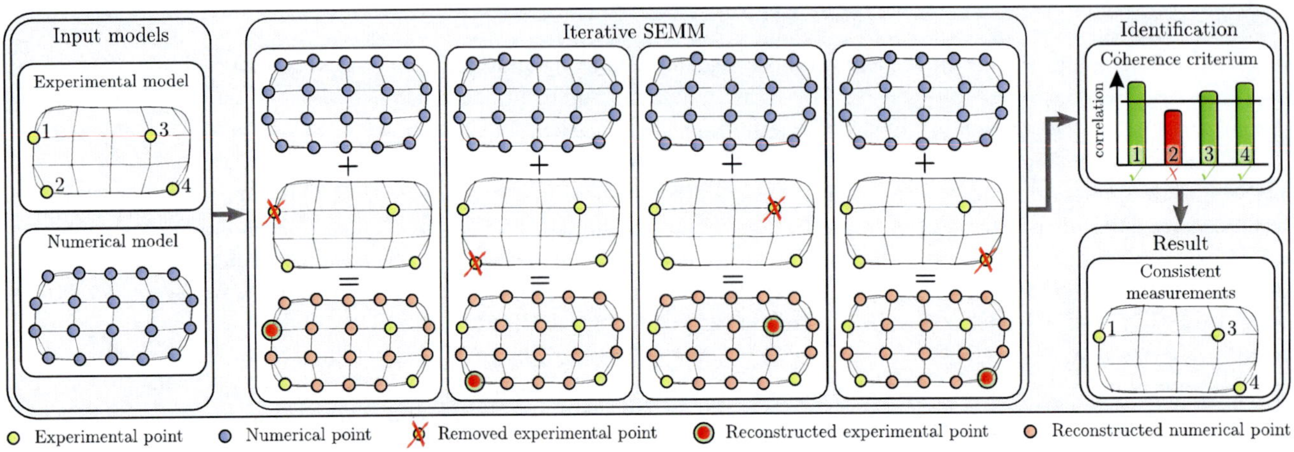

Fig. 8.2 Identification algorithm for the identification of inconsistent measurements in the frequency domain

8.2 System Equivalent Model Mixing—SEMM

System Equivalent Model Mixing (SEMM) was first introduced by Klaassen et al. [2]. The method forms a hybrid structural dynamic model by mixing the numerical and experimental FRFs. The method is based on the parent model (Fig. 8.1a), which provides the extensive DoF set and is usually of numerical nature. The dynamic properties are introduced by the overlay model (Fig. 8.1b), which is generally obtained by experiment. To form the final hybrid model (Fig. 8.1d) the dynamic properties of the parent model are eliminated with the removed model (Fig. 8.1c).

The compatibility and equilibrium conditions between equivalent models are imposed using the Lagrange Multiplier Frequency-Based Substructuring (LM FBS) framework [3], resulting in following form:

$$\mathbf{Y}^{\text{SEMM}} = [\mathbf{Y}]^{\text{par}} - \begin{bmatrix} \mathbf{Y}_{\text{ib}} \\ \mathbf{Y}_{\text{bb}} \end{bmatrix}^{\text{par}} \left(\mathbf{Y}^{\text{rem}}\right)^{-1} \left(\mathbf{Y}^{\text{rem}} - \mathbf{Y}^{\text{ov}}\right) \left(\mathbf{Y}^{\text{rem}}\right)^{-1} \left[\mathbf{Y}_{\text{bi}} \; \mathbf{Y}_{\text{bb}}\right]^{\text{par}}. \tag{8.1}$$

8.3 Identification of Inconsistent Measurements in the Frequency Domain

The identification algorithm is depicted in Fig. 8.2. Its operation requires an experimental and numerical model of analyzed structure. The latter represents only a tool for extending the DoFs within the SEMM method. In the second step, an individual measurement is extracted from the experimental model and reconstructed using the SEMM method based on the remaining measurements in the experimental model. When the procedure is repeated for all DoFs of the experimental model, the reconstructed measurements are compared with real measurements via the coherence criterion [4]. DoFs, where the value of average coherence does not reach a coherence limit, are identified as inconsistent measurements.

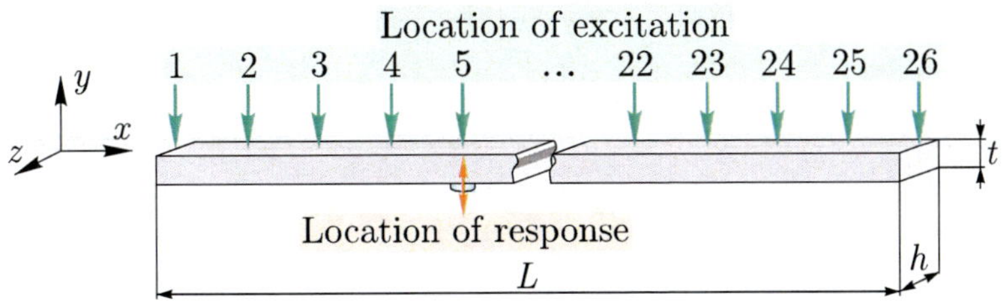

Fig. 8.3 Schematic presentation of beam A

Table 8.1 Parameters of beam A

Parameter	Value
L	300 mm
h	40 mm
t	12 mm
E	205 GPa
ρ	7820 kg/m^3

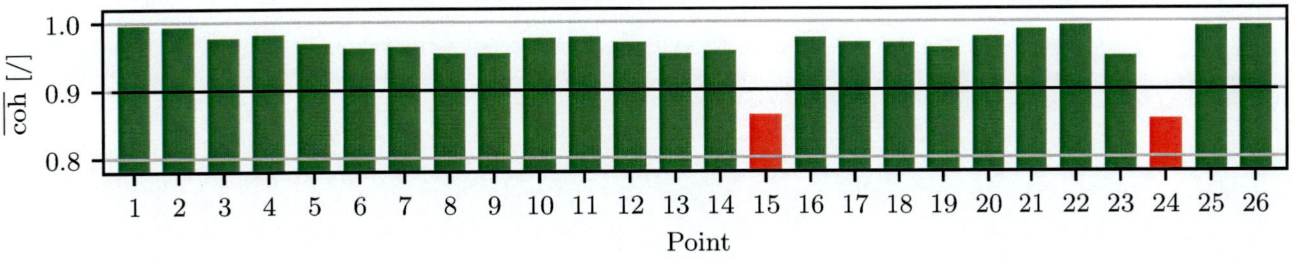

Fig. 8.4 Average value of the coherence and LAC criterion. (■) Consistent Measurement Coherence, (■) Inconsistent Measurement Coherence, (——) Boundary Coherence

8.4 Experimental Case Study

The applicability of the proposed method for the identification of inconsistent measurements is demonstrated on a simple beam structure with a rectangular cross-section. The beam A-A represents the main system assembled from two equal beams A (Fig. 8.3). In this paper, the dynamic response of the beam A-A will be obtained by coupling the dynamics of two identical beams A. On beam A, 26 equidistant measuring points were used (Fig. 8.3) for experimental and numerical model (Table 8.1).

The result of identification criteria is the coherence diagram, shown in Fig. 8.4, where the boundary-coherence value was chosen to be 0.90. The measurements at points 15 and 24 are identified as inconsistent and will be removed from the existing experimental data set.

To show the efficiency of the proposed method for the identification of inconsistent measurements, two identical beams A were coupled using the LM FBS method to form a dynamic model of the coupled beam A-A. To validate the coupling results, the reference experimental dynamic model of the coupled beam A-A was obtained independently. An example of the coupled structure's dynamic response is shown in Fig. 8.5. The beam was coupled using the entire experimental model as well as the experimental model, which contained only consistent measurements. The coupling results were significantly more accurate when we used the latter experimental model.

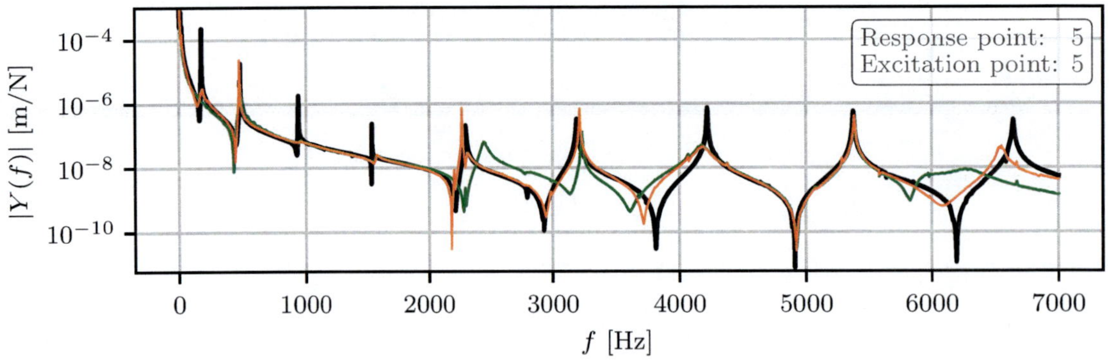

Fig. 8.5 FRF of coupled system. (—) Ref. Meas., (—) SEMM—consistent Meas., (—) SEMM—all Meas.

References

1. Chen, Y., Avitabile, P., Dodson, J.: Data consistency assessment function (DCAF). Mech. Syst. Signal Proces. **141**, 106688 (2020)
2. Klaassen, S.W., van der Seijs, M.V., de Klerk, D.: System equivalent model mixing. Mech. Syst. Signal Proces. **105**, 90–112 (2018)
3. de Klerk, D., Rixen, D.J., de Jong, J.: The frequency based substructuring (FBS) method reformulated according to the dual domain decomposition method. In: Proceedings of the 24th International Modal Analysis Conference, A Conference on Structural Dynamics, pp. 1–14 (2006)
4. van der Seijs, M.V., van den Bosch, D.D., Rixen, D.J., de Klerk, D.: An improved methodology for the virtual point transformation of measured frequency response functions in dynamic substructuring. In: 4th ECCOMAS Thematic Conference on Computational Methods in Structural Dynamics and Earthquake Engineering (COMPDYN), pp. 4334–4347 (2013)

Chapter 9
Weakening of the Interface Constraints in Modal Substructuring Using Singular Value Decomposition

Miha Pogačar, Tomaž Bregar, Gregor Čepon, and Miha Boltežar

Abstract An experimental approach to modal substructuring often proves to be problematic. One of the issues to be addressed is to determine a suitable set of constrain equations to be applied when considering an interface with multiple connection points. An approach to solving this problem was presented in the transmission simulator method (TSM), where a truncated set of mode shapes of a flexible fixture is used to project the constraints on the fixture's vector subspace. In this paper, the possibility of using the singular value decomposition (SVD) to obtain an appropriate subspace basis for weakening the constrain equations is explored. The most dominant response modes, acquired by performing SVD on the frequency response functions at the interface, are selected to weaken the constrain equations. The proposed approach is validated in a numerical case study.

Keywords Structural dynamics · Dynamic substructuring · Component mode synthesis · Weakening constraint equations · Singular value decomposition

9.1 Introduction

Component mode synthesis (CMS) is a well-established procedure in the field of numerical simulations, while several problems may arise in the case of an experimental approach to modal substructuring [1]. One of the possible issues is excessive stiffening, which is typical in the cases with multiple connection points when applying physical constraints. Typically, some of the physical constraints are redundant, which causes locking of the interface due to the inevitable measurement errors. An approach to solving this problem was presented in the transmission simulator method [2], where a truncated set of mode shapes of a flexible fixture is used to project the constraints on the fixture's vector subspace. A least squares fit provides weakening of the constraints and reduces the effect of measurement errors. However, a predicament may arise when selecting a proper subspace basis for projection, as the selection of base vectors and truncation limit can greatly affect the results of substructuring procedure.

9.2 Component Mode Synthesis—Primal Formulation

According to the primal CMS formulation [1, 2], the following linear equations of motion can be written in the coupling procedure of substructures A and B:

$$\begin{bmatrix} \mathbf{I}^{(\mathrm{A})} & \mathbf{0} \\ \mathbf{0} & \mathbf{I}^{(\mathrm{B})} \end{bmatrix} \begin{bmatrix} \ddot{\eta}^{(\mathrm{A})} \\ \ddot{\eta}^{(\mathrm{B})} \end{bmatrix} + \begin{bmatrix} \mathrm{diag}\left(\omega_r^{(\mathrm{A})2}\right) & \mathbf{0} \\ \mathbf{0} & \mathrm{diag}\left(\omega_r^{(\mathrm{B})2}\right) \end{bmatrix} \begin{bmatrix} \eta^{(\mathrm{A})} \\ \eta^{(\mathrm{B})} \end{bmatrix} = \begin{bmatrix} \mathbf{\Phi}^{(\mathrm{A})} & \mathbf{0} \\ \mathbf{0} & \mathbf{\Phi}^{(\mathrm{B})} \end{bmatrix} \begin{bmatrix} f^{(\mathrm{A})} \\ f^{(\mathrm{B})} \end{bmatrix}, \tag{9.1}$$

M. Pogačar (✉) · G. Čepon · M. Boltežar
Faculty of Mechanical Engineering, University of Ljubljana, Ljubljana, Slovenia
e-mail: miha.pogacar@fs.uni-lj.si; gregor.cepon@fs.uni-lj.si; miha.boltezar@fs.uni-lj.si

T. Bregar
Gorenje d.o.o., Velenje, Slovenia
e-mail: tomaz.bregar@gorenje.com

M. S. Allen et al. (eds.), *Dynamic Substructures, Volume 4*, Conference Proceedings of the Society for Experimental Mechanics Series, https://doi.org/10.1007/978-3-030-75910-0_9

where for the each substructure s, $\eta^{(s)}$ is a vector of modal coordinates and $\omega_r^{(s)2}$ and $\Phi^{(s)}$ are the corresponding eigenvalues and mass normalized mode shapes. The external force vector is denoted as $f^{(s)}$, whereas $\mathbf{I}$ is an identity matrix and $\ddot{[]}$ indicates the second time derivative. The damping contribution was omitted in the paper for the sake of clarity. Constrain equations can be written as

$$\mathbf{W}^{\dagger}\,\mathbf{B}\begin{bmatrix}\Phi^{(A)} & \mathbf{0}\\ \mathbf{0} & \Phi^{(B)}\end{bmatrix}\begin{bmatrix}\ddot{\eta}^{(A)}\\ \ddot{\eta}^{(B)}\end{bmatrix}=\mathbf{0}\,, \tag{9.2}$$

where $\mathbf{B}$ is a signed Boolean matrix and $\mathbf{W}$ is (generally) a non-square vector basis. Its generalized inverse (denoted as $[]^{\dagger}$) is used to weaken the physical constrains in the least square sense. In the transmission simulator method [2], a truncated set of mode shapes of a substructure is used to project the constraints on its vector subspace.

9.3 Singular Value Decomposition

Several approaches are established in the field of proper orthogonal decomposition, in order to obtain low dimensional approximations of high dimensional processes [3, 4]. In this paper, the possibility of using the singular value decomposition (SVD) to obtain an appropriate subspace basis for weakening the constrain equations is explored. The most dominant response modes, acquired by performing SVD on the frequency response functions [5] at the interface, are selected to weaken the constrain equations.

Let the $\mathbf{H}$ be an input matrix, containing the interface frequency response functions (FRFs). Its singular value decomposition can be written as

$$\mathbf{S}=\mathbf{U}\,\mathrm{diag}(s)\,\mathbf{V}^{\mathrm{H}}\,, \tag{9.3}$$

where $\mathbf{U}$ and $\mathbf{V}$ are orthonormal matrices of left and right singular vectors, respectively, and s is a vector of singular values. Two types of FRF matrices at the interface are considered as an input matrices $\mathbf{H}$. First, single input–multiple output (SIMO) matrix is used in the form, shown in Fig. 9.1a. In the second case, multiple input–multiple output (MIMO) matrix is used, concatenated as depicted in Fig. 9.1b.

In both cases considered, matrix $\mathbf{W}$ is constructed from left singular vectors, corresponding to the set of dominant singular values. Since the left singular vectors are orthonormal, a calculation of the generalized inverse in not needed in the process of the weakening of the constraints but a simple transposition operation can be performed instead.

9.4 Numerical Study

Numerical study was performed on a free–free supported T-shaped structure, coupled to its symmetrical equivalent, as shown in Fig. 9.2. In the reference coupled (AB) structure, a homogeneous joint is assumed, whereas the coupling procedure is performed in 20 connection points (equal to the 60 physical constraints), as depicted in Fig. 9.2a. Material properties in the analysis are consistent with an aluminum alloy with Young modulus of 70 GPa and density of 2700 kg/m^3.

In the modal transformation of substructures A and B, all the free–free mass normalized mode shapes up to 10 kHz (altogether 21 modes) were considered, simulating experimentally accessible range. It is known that free modes are typically not an optimal basis for substructuring predictions [2]; however, such an approach was chosen in order to provide clear representation. As a criterion of the quality of the coupling procedure, the accuracy of the coupled natural frequencies in

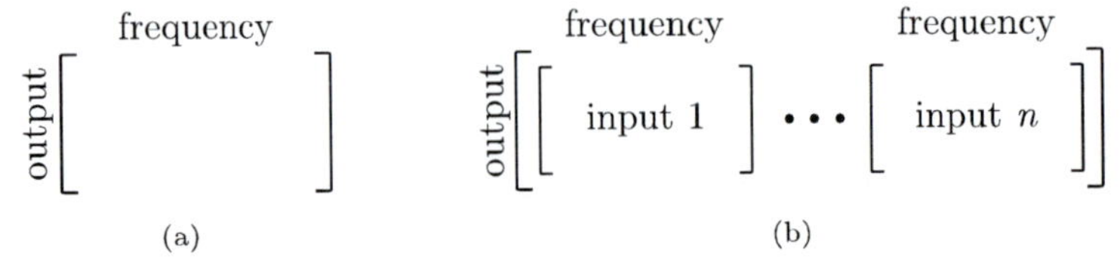

Fig. 9.1 Shape of the input FRF matrix: (**a**) SIMO; (**b**) MIMO

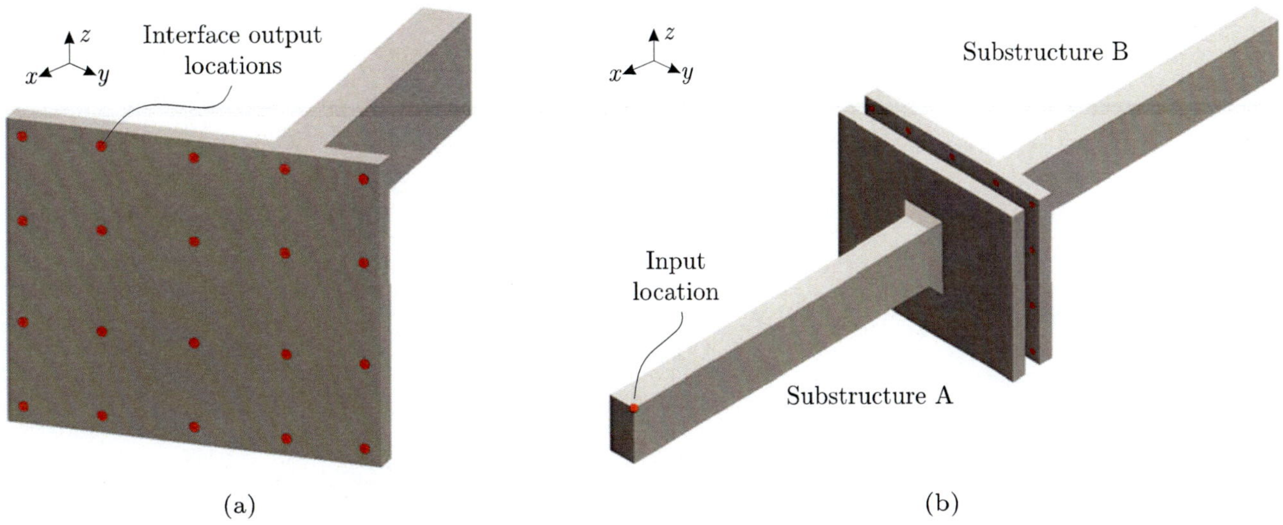

Fig. 9.2 (**a**) Connection points (denoted by red dots); (**b**) Depiction of coupling process of substructures A and B

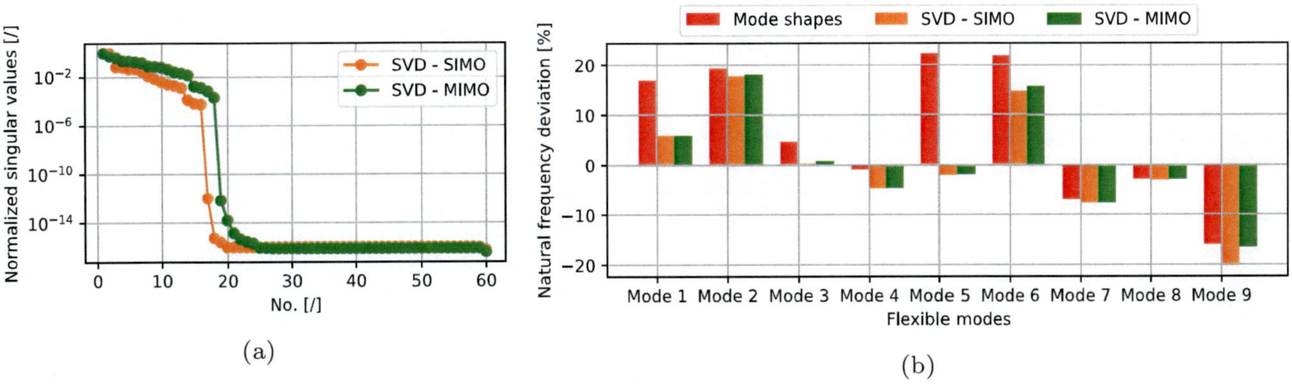

Fig. 9.3 (**a**) Singular values for SIMO (16 dominant values) and MIMO (18 dominant values) approach; (**b**) Coupled natural frequency predictions compared with the reference

the frequency range up to 4 kHz (9 flexible modes) and the coherence of the predicted FRF at the input location (marked in Fig. 9.2b) will be considered.

9.4.1 Results

In this section, three approaches are considered. In the first approach, all 21 modes are to be included in the weakening matrix **W**. In the second case, 16 left singular vectors are considered, corresponding to the dominant singular values (Fig. 9.3a), obtained by SVD of the SIMO FRF matrix (y-axis excitation at the input location and 60 responses at the interface output locations). Similarly, in the third case a MIMO FRF matrix is considered, where x, y, and z excitation directions were considered at the input location and 18 dominant left singular vectors (Fig. 9.3a) were included in the **W** matrix. In Fig. 9.3b, normalized natural frequencies of the coupled structure are compared with the reference values.

In Fig. 9.4, a coupled amplitude spectra for y-axis driving point FRF at the input location point is compared with the reference.

Based on the presented results we can conclude that left singular vectors of the interface FRF matrix can be used as an efficient basis to weaken the constraints in the coupling procedure. Compared with the established use of modal shapes, similar quality of the coupled system response can be achieved. In the future, the proposed procedure also requires experimental validation.

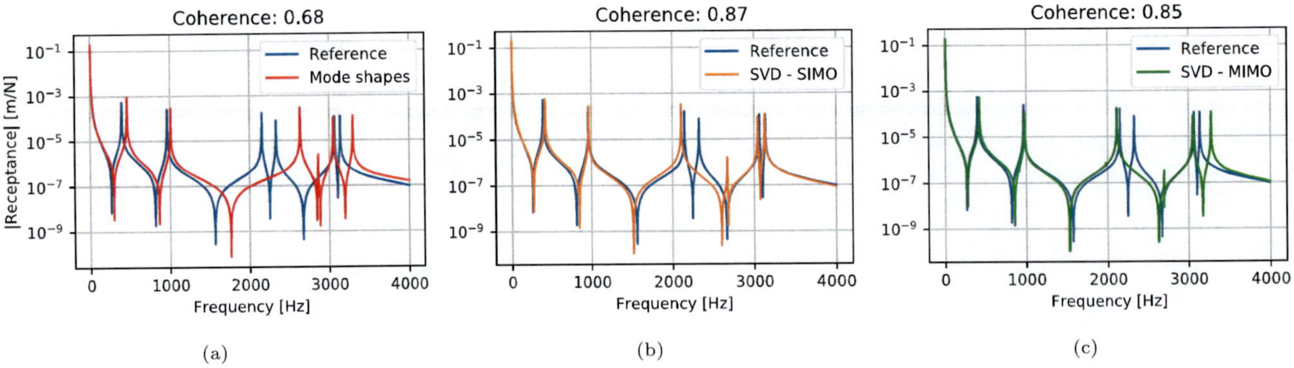

Fig. 9.4 Comparison of the coupled and reference FRFs at the input location: (**a**) Mode shapes; (**b**) SIMO; (**c**) MIMO

References

1. Klerk, D., Rixen, D., Voormeeren, S.: General framework for dynamic substructuring: history, review, and classification of techniques. AIAA J. **46**, 1169–1181 (2008)
2. Allen, M.S., Mayes, R.L., Bergman, E.J.: Experimental modal substructuring to couple and uncouple substructures with flexible fixtures and multi-point connections. J. Sound Vib. **329**(23), 4891–4906 (2010)
3. Wu, C., Liang, Y., Lin, W., Lee, H., Lim, S.: A note on equivalence of proper orthogonal decomposition methods. J. Sound Vib. **265**, 1103–1110 (2003)
4. Kerschen, G., Golinval, J.-C.: Physical interpretation of the proper orthogonal modes using the singular value decomposition. J. Sound Vib. **249**(5), 849–865 (2002)
5. Allemang, R.J., Phillips, A.W., Allemang, M.R.: Application of principal component analysis methods to experimental structural dynamics. In: Structural Dynamics, vol. 3, pp. 477–498. Springer, Berlin (2011)

Chapter 10
Introducing pyFBS: An Open-Source Python Package for Frequency Based Substructuring and Transfer Path Analysis

Tomaž Bregar, Ahmed El Mahmoudi, Miha Kodrič, Gregor Čepon, Miha Boltežar, and Daniel J. Rixen

Abstract pyFBS is an open-source Python package for Frequency Based Substructuring. The package implements an object-oriented approach for dynamic substructuring. State-of-the-art methodologies in frequency based substructuring, such as virtual point transformation and system equivalent model mixing, are available within the pyFBS. Each method can be used as a standalone or interchangeably with others. Tools are provided to easily visualize components and configure the measurement setup. Also operational deflection shapes and mode shapes can be animated directly within the 3D display. Furthermore, basic and application examples are available, together with numerical and experimental datasets, to enable the user to get familiar with the work flow of the package. This paper showcases the use of the pyFBS on two example structures. Firstly, a simple beam-like structure is used to depict the use of the 3D display, FRF synthetization, virtual point transformation and system equivalent model mixing. Secondly, an automotive test-structure is used to show the use of the pyFBS on real-life complex structure, where the in-situ transfer path analysis is used to characterize blocked forces. The development of the pyFBS is an ongoing effort, as it is actively being used as a research tool. Additional features and new methods will be integrated within the pyFBS in the near future.

Keywords Frequency based substructuring · Transfer path analysis · Virtual point transformation · System equivalent model mixing · Python · Open-source

10.1 Introduction

Open-source software has seen an increased popularity both in research and industry practice. In academic research, open-source approach provides an additional transparency, since the open-source code is constantly peer reviewed. Furthermore, by sharing and collaborating on similar topics a higher number of leading experts with a shared interest can get involved in the development. Commonly, open-source software uses a modular approach, where some parts of the code can be used independently, making it easy to connect various packages and maintain the compatibility.

This paradigm of divide and conquer is not only found in modular software but is also a powerful strategy in science, engineering and complex technological problems that are often decomposed into smaller, simpler subsystems, which are more easily handled. With this, the complexity of the overall problem is reduced and an invaluable insight is obtained into the optimization and troubleshooting of each individual component. The subsystems can also be assembled back together and with that the system can be analysed as a whole.

Dynamic Substructuring (DS) is an engineering concept where dynamic systems are characterized in terms of their subcomponents. There are several different ways of formulating the dynamics of substructures. One of them is with Frequency Response Functions (FRFs), which describe the response as the result of a unit harmonic force. Modelling of structures with FRFs is primarily used with experimental methods, where substructures can be modelled entirely through

T. Bregar (✉)
Gorenje d.o.o., Velenje, Slovenia
e-mail: tomaz.bregar@gorenje.com

M. Kodrič · G. Čepon · M. Boltežar
Faculty of Mechanical Engineering, University of Ljubljana, Ljubljana, Slovenia

A. E. Mahmoudi · D. J. Rixen
Technische Universität München, Institute for Applied Mechanics, Garching, Germany

M. S. Allen et al. (eds.), *Dynamic Substructures, Volume 4*, Conference Proceedings of the Society for Experimental Mechanics Series, https://doi.org/10.1007/978-3-030-75910-0_10

an experimental approach. Various methods were already investigated in the 1970s [1] and 1980s (e.g. [2–5]). Due to complicated formulations and difficulties in obtaining good measurements, the method was hardly applicable.

In recent years, the Frequency Based Substructuring (FBS) has gained an increased popularity, primarily due to the proper formulation of the problem and better measurement hardware [6–8]. Within the FBS, it is also possible to build hybrid models, in which experimentally characterized and numerically modelled parts can be combined [9].

This paper shows the capabilities of the pyFBS: an open-source Python package for frequency based substructuring and transfer path analysis. Current state-of-the-art methods in FBS are implemented within the pyFBS, such as virtual point transformation [10], singular vector transformation [11] and system equivalent model mixing [9]. Additional, numerical and experimental datasets of two test-structures are provided to evaluate currently implemented features on real examples.

Note that a commercial software for TPA and FBS successfully used in industries over the last years is available from Vibes.Technology. Although the methodologies are similar, the open-source project presented in this paper and initiated in the scientific community is however not directly related to the Vibes software.

10.2 Features of pyFBS

The pyFBS introduces an object-oriented approach for dynamic substructuring in frequency domain. Current state-of-the-art FBS methodologies are implemented within the package. Each method can be used interchangeably with others or as a standalone unit. In the next sections, the key features of the pyFBS are outlined.

10.2.1 Test-Structures

Within the pyFBS numerical and experimental datasets of two test-structures are available, which can be directly downloaded and used. The first AM test-structure is an academic example of a beam-like structure, which is depicted in Fig. 10.1a.[1] The two components of this AM structure are machined from a single aluminium block. The system is lightly damped and both single point and line contact can be made with the two components. The second structure is an automotive test-structure designed to represent an engine block suspended in its frame, depicted in Fig. 10.1b. In this lab mock-up, a massive steel block connected to an electrodynamic shaker represents the engine block and is mounted through three rubber mounts (engine, roll and transmission mount) to the receiver structure. Both test-structures are intended to be used to evaluate different concepts of dynamic substructuring and transfer path analysis in a laboratory environment.

10.2.2 3D Display

Substructures and positions of impacts, sensors and channels can be visualized in 3D display. For the visualization a Python package, PyVista [13] is used, which enables an intuitive and streamlined way to display and animate data in the 3D space. Following the elegant interaction concept developed in [14] and later optimized by Vibes.Technology, both sensors and impacts can be interactively positioned on the substructures and the updated positions, as well as orientations, can be directly used with other features. An example of use of the 3D display is shown in Fig. 10.2.

10.2.3 FRF Synthetization

Frequency Response Functions can be synthesized with mode superposition from a modal model. The modal model can be calculated from mass and stiffness matrices or directly imported using the pyansys package [15]. In Fig. 10.3, the 1st and 4th mode shape of the AM test-structure is depicted. The mass and stiffness matrices can also be derived analytically (for a discrete mass spring system) and imported to calculate the FRFs. Damping can be introduced as a modal damping for each

[1] This AM structure was inspired by the AB structure used in [12] to investigate new coupling strategies.

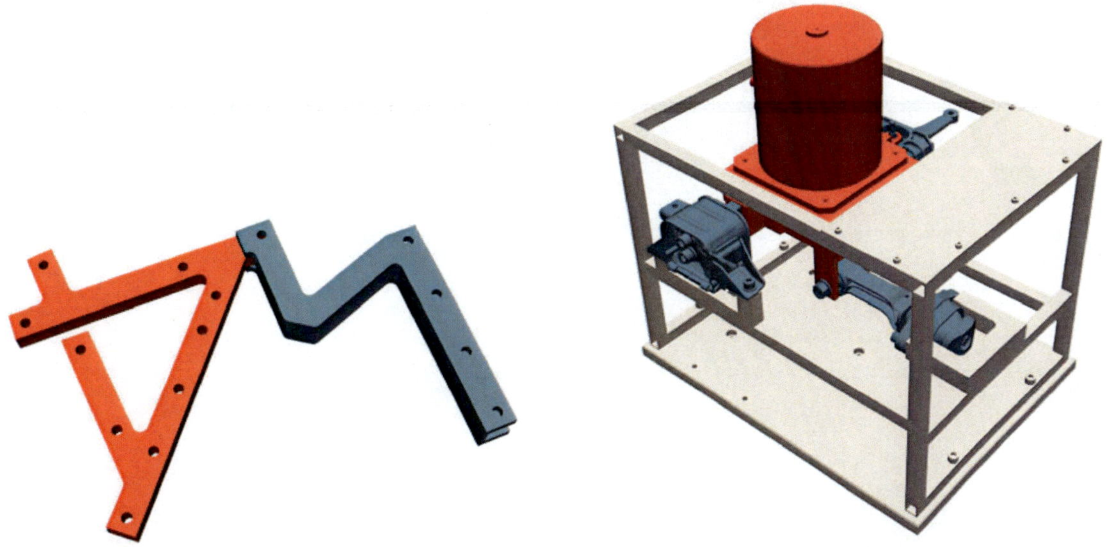

(a) (b)

Fig. 10.1 Two test-structures currently available within the pyFBS: (**a**) AM test-structure; (**b**) automotive test-structure

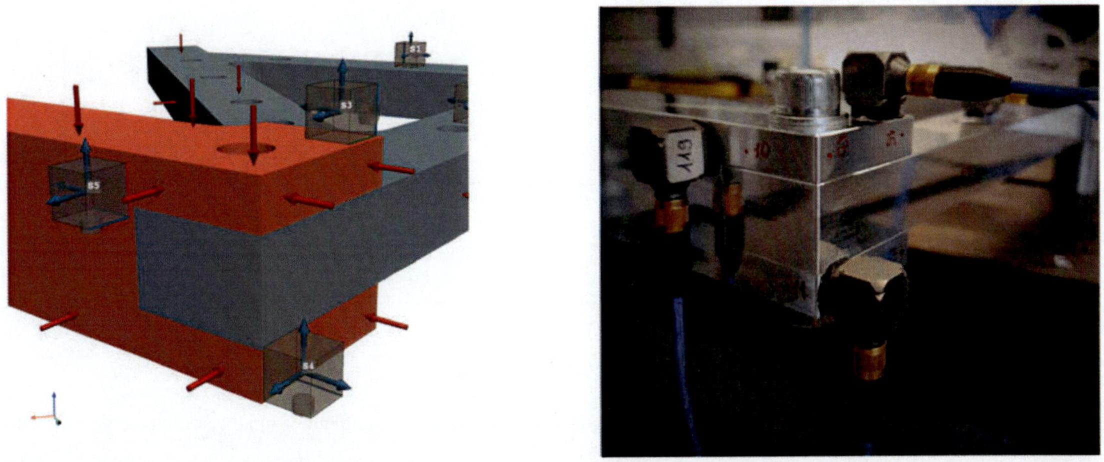

(a) (b)

Fig. 10.2 Placement of sensors around a point interface: (**a**) virtual setup; (**b**) experimental setup

mode shape. Additionally, random noise can be added to the FRFs to obtain a realistic response, representing experimental measurements. In Fig. 10.4, a synthesized accelerance FRF is depicted with different number of mode shapes and an example of added random noise.

10.2.4 Virtual Point Transformation

Virtual point transformation projects measured sensor displacements and force inputs around the interface on the predetermined interface deformation modes [10, 14]. If the interface is considered to be locally rigid, then 6 virtual degrees of freedom are assumed for each virtual point (3 translations and 3 rotations). The transformation is achieved through displacement- and force-transformation matrices, which are defined based on the relative position of sensors to the virtual point.

Within the pyFBS, the transformation matrices can be defined based on the positions and orientations from the 3D display. An example of transformed virtual FRFs is shown in Fig. 10.5 (the transformation is applied on the interface shown in

Fig. 10.3 Numerical mode shapes in the assembled configuration AB: (**a**) 1st mode shape; (**b**) 4th mode shape

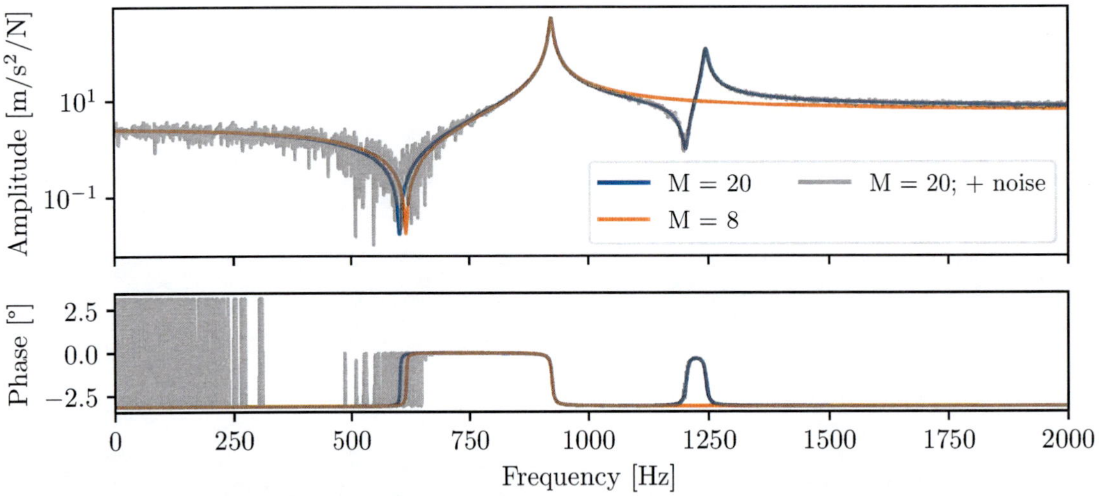

Fig. 10.4 Synthesized driving-point frequency response function on the A structure (M denotes number of modes used in the synthetization)

Fig. 10.2). One of the important features of the VPT is also the measurement-quality indicators. They enable a user to assess the overall quality of the measurement. The indicators are defined by comparing filtered response (by back-projection to original degrees of freedom) with the measured response [14]. An example of the specific impact and channel indicators is shown in Fig. 10.6.

An expanded VPT is also supported, where directly measured rotational responses can be integrated in the transformation [16].

10.2.5 System Equivalent Model Mixing

System equivalent model mixing is a method with which multiple equivalent response models can be mixed using the LM FBS [9]. A schematic representation of SEMM is schematically depicted in Fig. 10.7. The overlay model provides the dynamic properties and the parent model a DoF-set. The dynamic properties of the parent model (removed model) are, at the end, removed from the final hybrid model.

The hybrid SEMM model is generally formed by mixing a numerical with an experimental model. A numerical model is commonly used as a parent model, providing a full DoF set and an experimental model is used as an overlay model, providing the experimental properties. An application of SEMM on the AM structure using the pyFBS is shown in Fig. 10.8.

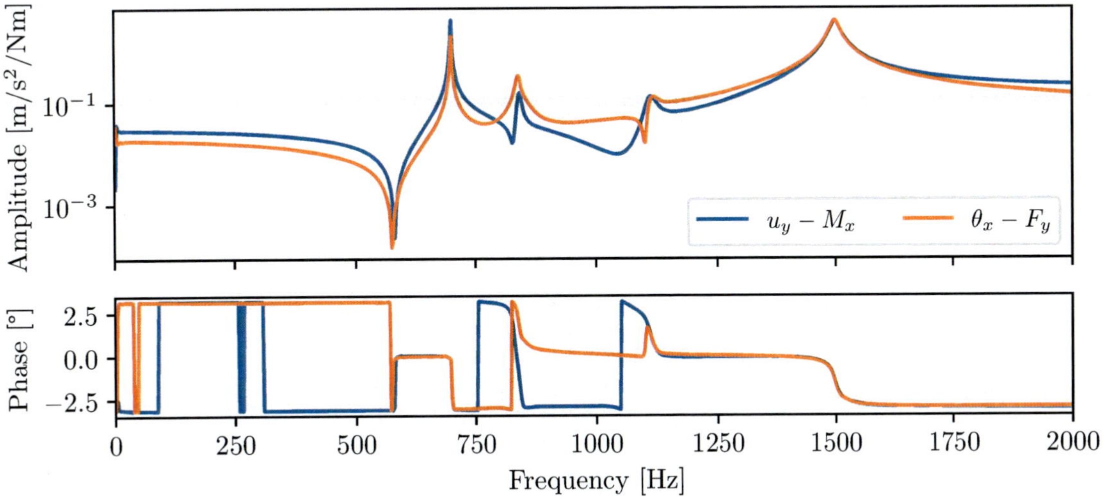

Fig. 10.5 Reciprocity check of virtual point FRFs at a point interface of the AM structure

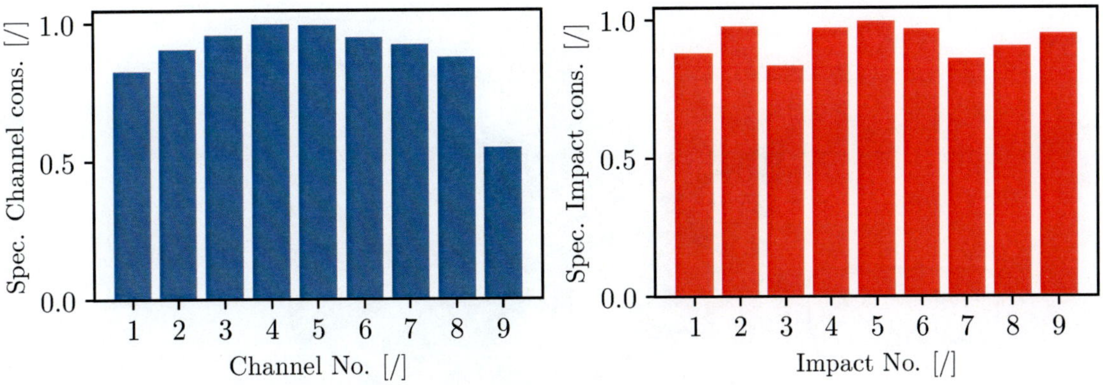

Fig. 10.6 Specific impact and channel consistency of the VPT applied on the AM structure

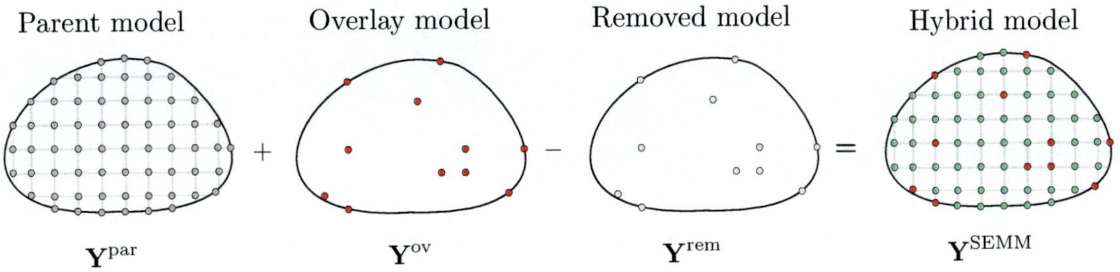

Fig. 10.7 Schematic representation of System Equivalent Model Mixing (SEMM) [9]

10.3 Source Characterization with the In-situ TPA

With the in-situ transfer path analysis, the equivalent forces can be determined directly from the assembled configuration. The equivalent forces are a sole property of the source [7], which makes them indispensable in the design process, since various modifications of the passive side can be evaluated, without the need to repeat the source characterization. It is a common industry practice that the active components are outsourced to external suppliers. By utilizing the property of equivalent forces, the target settings can be specified on a component level and thus outsourcing the task of source characterization.

Here an application of in-situ TPA is shown on an automotive laboratory test-structure (Fig. 10.9). The test-structure is designed to mimic the dynamics in a real car of an engine-transmission-unit flexibly mounted on a chassis. Rubber mounts from the automotive industry are used to suspend a steel plate holding the excitation source, which is an electrodynamic

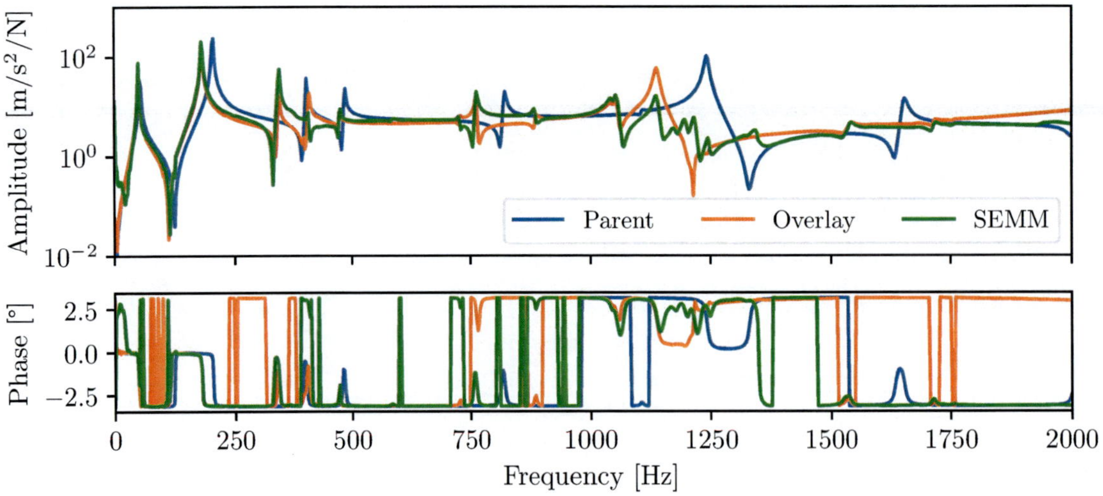

Fig. 10.8 Comparison of experimental measurement, numerical model and final hybrid model on the AM structure

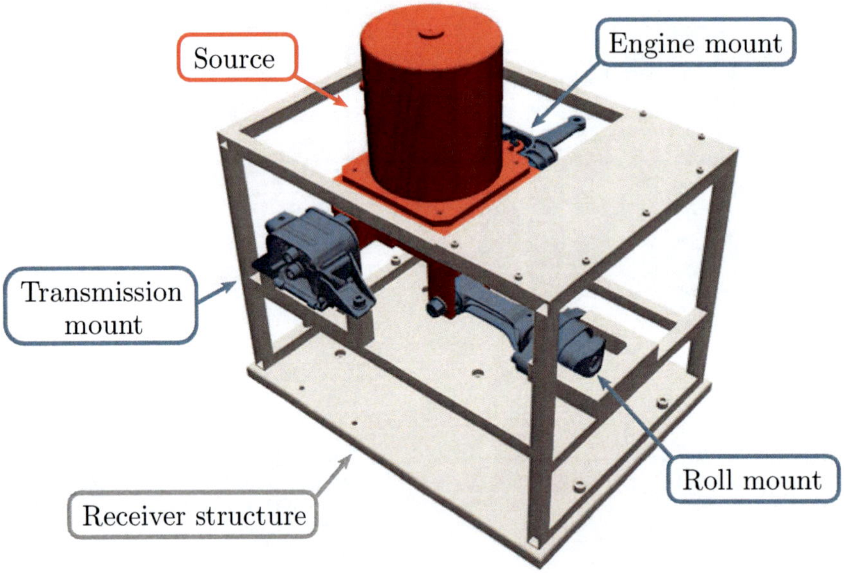

Fig. 10.9 Schematic representation of the automotive laboratory test-structure [17]

shaker. The structure is intended as a laboratory test bench for an application of different concepts of dynamic substructuring and transfer path analysis [17].

To properly describe the interface between the source structure and each rubber mount a set of virtual forces has to be defined. All three interfaces can be considered locally rigid; therefore, virtual point transformation can be used to define a set of 6 virtual forces (3 forces and 3 moments). Locations of the VP are depicted on Fig. 10.10. With the use of virtual point transformation, the interface problem is weakened and the measurement errors are to some extent averaged out. For each virtual point 3 tri-axial accelerometers (9 channels) were used and 9 excitation points; therefore, altogether 27 excitation points were used for 18 virtual point forces/moments.

10.3.1 Equivalent Forces

After performing FRF measurement, applying virtual point transformation and operational measurements, the equivalent forces can be calculated. For the operational response, the system is commonly operating under various loads and different

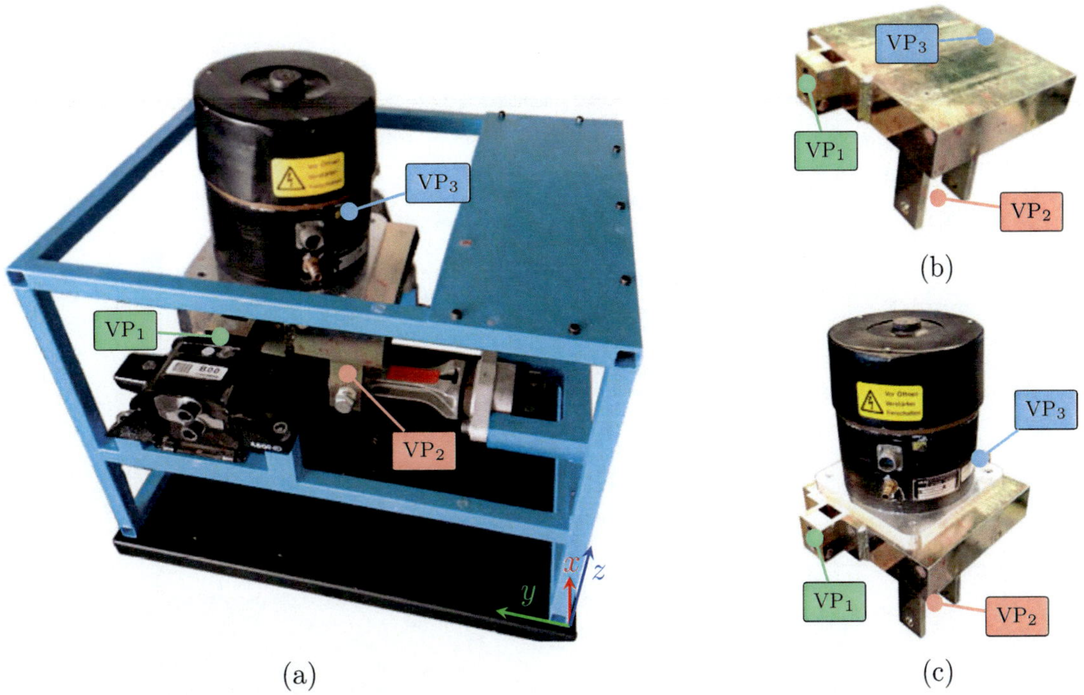

Fig. 10.10 Test-structure used for the comparison: (**a**) assembly $\mathbf{Y}^{\mathrm{AB}}$ representing an engine-transmission unit suspension [18]; (**b**) transmission simulator $\mathbf{Y}^{\mathrm{TS}}$ used for additional measurement space; (**c**) source $\mathbf{Y}^{\mathrm{A}}$ representing the car engine

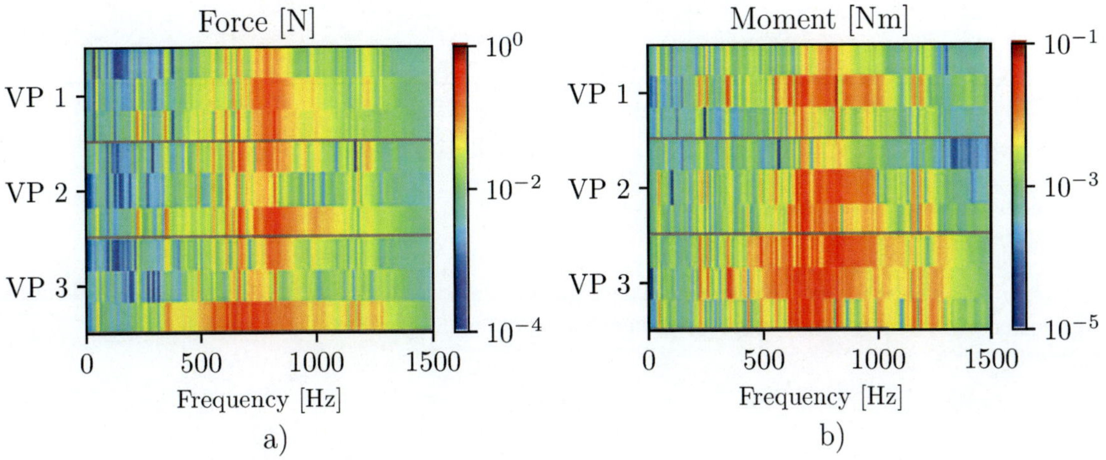

Fig. 10.11 Identified equivalent force: (**a**) virtual forces; (**b**) virtual moments

speed settings; however, here an additional impact excitation location on the source structure is used as an operational excitation (see [17] for details). In Fig. 10.11, both identified equivalent forces and moments at each virtual point are shown.

After the equivalent forces are determined, the partial responses from each virtual point force or moment can be calculated. In Fig. 10.12, partial responses are depicted on a heat-map.

10.3.2 On-Board Validation

On-board validation compares the operational response with the estimated response from the determined equivalent forces: the estimated response and the data collected for the blocked force being taken from the same assembly. With this, one can

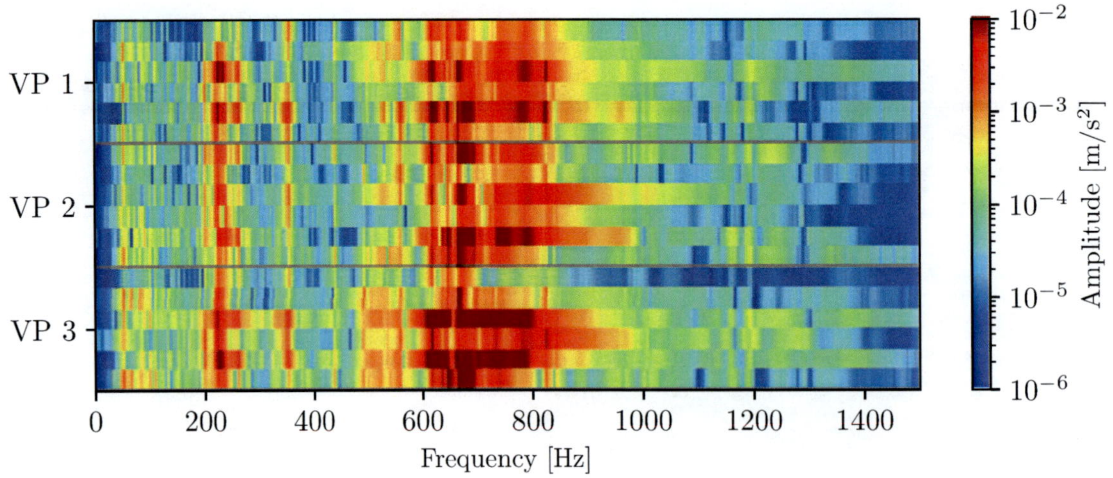

Fig. 10.12 Reconstructed partial responses at a single reference location based on identified equivalent forces

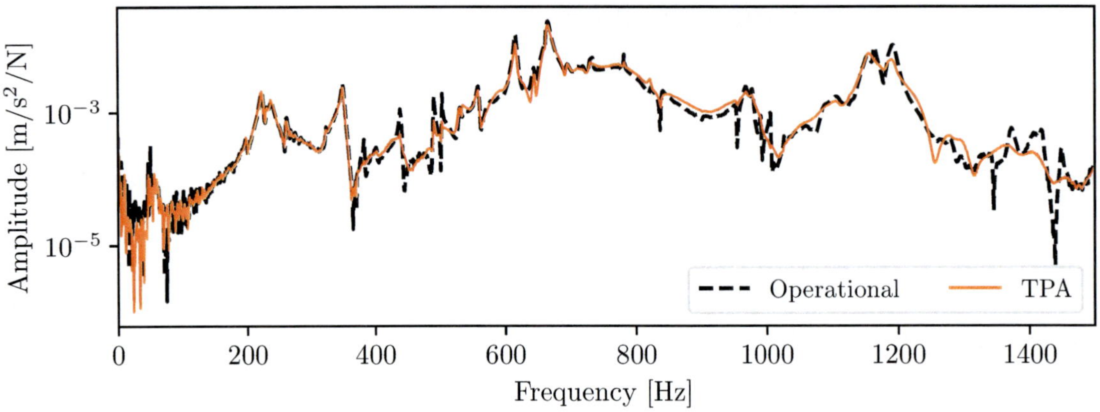

Fig. 10.13 On-board validation of the impact excitation as an operational excitation

verify if the calculated equivalent forces can completely describe the source activity. In Fig. 10.13, an on-board validation is provided for estimated equivalent forces. A good agreement can be observed over the whole frequency range.

10.4 Conclusion

The development of open-source software has gained popularity in the past decade in the scientific community to allowing the scientific community to share their work and interact efficiently. New packages are introduced by the community on a daily basis. With an open-source approach to scientific research, additional transparency and reliability can be obtained. The pyFBS is an open-source Python package for Frequency Based Substructuring and Transfer Path Analysis. The development of the pyFBS is an ongoing effort and is currently being used as a research tool. The modular concept enables the user to exploit only the required modules. It supports current state-of-the-art methodologies in frequency based substructuring, such as virtual point transformation, singular vector transformation and system equivalent model mixing. More applications on the topic of dynamic substructuring and of Transfer Path Analysis will be introduced in the documentation. Furthermore, methods such as operational source identification will get integrated in the pyFBS in the near future.

The source-code of pyFBS is available on https://gitlab.com/pyFBS and an extensive documentation is hosted on https://pyfbs.readthedocs.io.

Acknowledgments The pyFBS package was developed as a part of collaboration between the Laboratory for Dynamics of Machines and Structures (LADISK), Faculty of Mechanical Engineering, University of Ljubljana (UL FME) and the Chair of Applied Mechanics (AM), Technical University of Munich (TUM).

References

1. Klosterman, A.L.: A combined experimental and analytical procedure for improving automotive system dynamics, Ph.D. Thesis, University of Cincinnati, Department of Mechanical Engineering (1971)
2. Martinez, D.R., Carrie, T.G., Gregory, D.L., Miller, A.K.: Combined Experimental/Analytical Modelling using component modes synthesis. In: 25th Structures, Structural Dynamics and Materials Conference, Palm Springs, pp. 140–152 (1984)
3. Crowley, J.R., Klosterman, A.L., Rocklin, G.T., Vold, H.: Direct structural modification using frequency response functions. In: Proceedings of IMAC II (1984)
4. Jetmundsen, B., Bielawa, R.L., Flannelly, W.G.: Generalized frequency domain substructure synthesis. J. Am. Helicopter Soc. **33**(1), 55–64 (1988)
5. Urgueira, A.P.V.: Dynamic analysis of coupled structures using experimental data. Ph.D. Thesis, Imperial College, London (1989)
6. de Klerk, D., Rixen, D.J., Voormeeren, S.N.: General framework for dynamic substructuring: history, review and classification of techniques. AIAA J. **46**, 1169–1181 (2008)
7. van der Seijs, M.V., de Klerk, D., Rixen, D.J.: General framework for transfer path analysis: history, theory and classification of techniques. Mech. Syst. Signal Proces. **68–69**, 217–244 (2016)
8. Rixen, D., Godeby, T., Pagnacco, E.: Dual assembly of substructures and the FBS method: application to the dynamic testing of a guitar. In: International Conference on Noise and Vibration Engineering, ISMA (2006)
9. Klaassen, S.W.B., der Seijs, M.V., de Klerk, D.: System equivalent model mixing. Mech. Syst. Signal Proces. **105**, 90–112 (2018)
10. de Klerk, D., Rixen, D.J., Voormeeren, S.N., Pasteuning, F.: Solving the RDoF problem in experimental dynamic substructuring. In: Proceedings of the 26th International Modal Analysis Conference, A Conference on Structural Dynamics (2010)
11. Trainotti, F., Bregar, T., Klaassen, S., Rixen, D.J.: Experimental decoupling of substructures by singular vector transformation. Mech. Syst. Signal Proces. (2020)
12. van der Seijs, M.V., Pasma, E.A., van den Bosch, D.D., Wernsen, M.W.F.: A benchmark structure for validation of experimental substructuring, transfer path analysis and source characterisation techniques. In: Dynamics of Coupled Structures, vol. 4. Springer International Publishing, Cham, pp. 295–305 (2017)
13. Sullivan, C.B., Kaszynski, A., PyVista: 3d plotting and mesh analysis through a streamlined interface for the visualization toolkit (VTK). J. Open Source Softw. **4**(37), 1450 (2019)
14. van der Seijs, M.V.: Experimental Dynamic Substructuring: Analysis and design strategies for vehicle development. Ph.D. Thesis, TU Delft (2016)
15. Kaszynski, A.: pyansys: Python Interface to MAPDL and Associated Binary and ASCII Files (2020). https://doi.org/10.5281/zenodo.4009467
16. Bregar, T., Holeček, N., Čepon, G., Rixen, D.J., Boltežar, M.: Including directly measured rotations in the virtual point transformation. Mech. Syst. Signal Proces. **141**, 106440 (2020)
17. El Mahmoudi, A., Trainotti, F., Park, K., Rixen, D.J.: In-situ TPA for NVH analysis of powertrains: an evaluation on an experimental test setup. In: AAC 2019: Aachen Acoustics Colloquium/Aachener Akustik Kolloquium, Aachener Akustik Kolloquium 2019, Aachen, 25 Nov 2019–27 Nov 2019, pp. 1–9 (2019)
18. Bregar, T., Mahmoudi, A.E., Čepon, G., Rixen, D.J., Boltežar, M.: Performance of the expanded virtual point transformation on a complex test structure. Exp. Tech. **45**, 83–93 (2021)

Chapter 11
Analysis of Friction Induced Mode Coupling Instabilities Using Dynamic Substructuring

Jacopo Brunetti, Walter D'Ambrogio, and Annalisa Fregolent

Abstract In complex vibrating systems, contact and friction forces can produce a dynamic response of the system (friction induced vibrations). They can arise when different parts of the system move one with respect to the other generating friction force at the contact interface. Component mode synthesis and more in general substructuring techniques represent a useful and widespread tool to investigate the dynamic behavior of complex systems, but classical techniques require that the component subsystems and the coupling conditions (compatibility of displacements and equilibrium of forces) are time invariant. In previous papers, it was shown that contact problems can be cast in the framework of dynamic substructuring by considering the models of the component substructures as time invariant, while the coupling conditions must be time dependent. In this paper a substructuring method is proposed that, depending on the contact assumption, is able either to account only for the macroscopic sliding between substructures, or to consider also the local vibrations of the contact points or to consider also the geometric nonlinearity due to the elastic deformation. This allows to adapt the contact algorithm to the contact problem that must be tackled, i.e. position dependent dynamics or friction induced vibrations.

Keywords Dynamic substructuring · Time-variant interface · Sliding contact · Friction · Friction induced vibrations

11.1 Introduction

Dynamic substructuring [1–3] allows to predict the dynamic behavior of a complex mechanical system, knowing the dynamics of its components [4–6]. Although classical substructuring techniques assume that the system is linear and time invariant, several attempts have been made newly in order to tackle also nonlinear and time-variant problems. In fact, on one hand the extension of substructuring techniques to include nonlinearities [7–9] represents a very promising approach, for example to account for the nonlinear localized effects introduced by joints [10] and more in general by connections [11, 12]. On the other hand, substructuring approach can be extended to time-variant systems[13, 14], for example to deal with time-variant systems composed of time-invariant substructures subjected to time-variant coupling conditions [15, 16]. This allows to tackle structures with a configuration dependent dynamics and more specifically, systems with sliding contact interfaces.

In this paper, a substructuring method is proposed that, depending on the contact assumption, is able either to account only for the macroscopic sliding between substructures [16, 17], or to consider also the local vibrations of the contact points, or to consider also the geometric nonlinearity due to the elastic deformation [18, 19]. The approach presented in [16] assumed that the time-variant compatibility and equilibrium conditions, arising from sliding contact, are a priori known and depend only on the kinematic boundary conditions. This assumption does not allow to account for the relative displacement at the contact interface due to the system deformation, that is relevant for dynamic contact problems [20–22]. In fact, it is the main cause of friction induced vibrations phenomena [23] such as dynamic instabilities [24, 25], stick-slip [26–28], or sprag-slip [29, 30]. For this reason, a more realistic contact assumption is also considered. It accounts for the deformation of the contacting bodies in the definition of the time-variant compatibility and equilibrium conditions, but neglects the variation of the normal

J. Brunetti (✉) · W. D'Ambrogio
Dipartimento di Ingegneria Industriale e dell'Informazione e di Economia, Università dell'Aquila, L'Aquila, Italy
e-mail: jacopo.brunetti@univaq.it; walter.dambrogio@univaq.it

A. Fregolent
Dipartimento di Ingegneria Meccanica e Aerospaziale, Università di Roma La Sapienza, Rome, Italy
e-mail: annalisa.fregolent@uniroma1.it

M. S. Allen et al. (eds.), *Dynamic Substructures, Volume 4*, Conference Proceedings of the Society for Experimental Mechanics Series, https://doi.org/10.1007/978-3-030-75910-0_11

force direction due to the deformation. Moreover, a true contact algorithm is developed in [19] that accounts for all the implication of body deformation, providing a more reliable definition of compatibility and equilibrium conditions.

Substructuring based methods are applied here to evaluate the response associated with unstable friction induced vibrations of a beam on beam system [31]. In particular, the results obtained using the different contact assumptions are compared with the truth solution provided by a well assessed numerical method [32] in order to highlight possible strengths and weaknesses of each formulation.

11.2 Substructuring with Sliding Friction Interfaces

Mechanical systems composed of n connected subsystems can be tackled very effectively using dynamic substructuring. This approach allows to model separately each component substructure and obtain the dynamic behavior of the assembled structure by considering the displacement compatibility and the force equilibrium conditions at the coupling interfaces. Generally, the compatibility and equilibrium conditions apply on the same set of coupling DoFs (collocated interfaces) [1]. However, to account also for sliding friction at the coupling interface the compatibility and the equilibrium conditions must be defined on different set of DoFs (non collocated interfaces)[16].

Therefore, using the substructuring approach, the overall system of equations representing the coupled system is composed of the equations of motion of the independent subsystems, the compatibility conditions, and equilibrium conditions.

$$\begin{cases} [M]\{\ddot{u}(t)\} + [C]\{\dot{u}(t)\} + [K]\{u(t)\} = \{f(t)\} + \{g(t)\} \\ [B_C(t)]\{u(t)\} = \{0\} \\ [L_E(t)]^T\{g(t)\} = \{0\} \end{cases} \tag{11.1}$$

These equations, referred to as the Three Field Formulation in the substructuring literature, express the equation of motion in the physical components $\{u(t)\}$ of the assembled structure. $[M]$, $[C]$, and $[K]$ are the mass, damping, and stiffness matrices respectively, which gather, in block diagonal format, the matrices of the component substructures; $\{f(t)\}$ is the vector of the external forces; $\{g(t)\}$ is the vector of the connecting forces. The matrix $[B_C(t)]$ expresses the compatibility between pairs of matching DoFs at time (t), whereas the $[L_E(t)]$ expresses the equilibrium of connecting forces at pair of matching DoFs at time (t). In case of relative sliding among substructures, matrices $[B_C(t)]$ and $[L_E(t)]$ become time dependent. Moreover, when sliding friction occurs at the contact interfaces, the equilibrium condition applies also along the DoFs in the tangential directions, which are not subject to any compatibility condition.

11.2.1 Dual Assembly

In dual assembly, each interface DoF is considered as many times as there are substructures connected through that DoF. The equilibrium condition $g_l^{(r)}(t) + g_m^{(s)}(t) = 0$ at a pair of interface DoFs can be ensured by setting $g_l^{(r)}(t) = -\lambda(t)$ and $g_m^{(s)}(t) = \lambda(t)$. Therefore, all the connecting forces $\{g(t)\}$ can be expressed by using Lagrange multipliers $\{\lambda(t)\}$ that represent the connecting force intensities:

$$\{g(t)\} = -[B_E(t)]^T\{\lambda(t)\}. \tag{11.2}$$

Since the equilibrium condition in the three field formulation (third equation of (11.1)) is always satisfied for any set of connecting forces intensities $\{\lambda(t)\}$, then:

$$[L_E(t)]^T[B_E(t)]^T = [0]. \tag{11.3}$$

Matrix $[B_E(t)]$ is generally different from the matrix $[B_C(t)]$ used in the three field formulation to express the compatibility. In fact, at the friction interface, contact forces act along the normal and the tangential directions, whereas the compatibility of displacements is defined only along the normal direction. Thus, the system of equation becomes

$$\begin{cases} [M]\{\ddot{u}(t)\} + [C]\{\dot{u}(t)\} + [K]\{u(t)\} + [B_E(t)]^T\{\lambda(t)\} = \{f(t)\} \\ [B_C(t)]\{u(t)\} = \{0\}. \end{cases} \tag{11.4}$$

11.2.2 Nonlinear Forward Increment Dual Assembly

This formulation relates the constraints in the deformed position at time $t_{n+1} = t_n + h$, where h is the integration time step, with Lagrange multipliers at time t_n.

$$\begin{cases} [M]\{\ddot{u}\}_n + [C]\{\dot{u}\}_n + [K]\{u\}_n + [B_E]^T_{n+1}\{\lambda\}_n = \{f\}_n \\ [B_C]_{n+1}\{u\}_{n+1} = \{0\} \end{cases} \tag{11.5}$$

Note that the matrices $[B_C]$ and $[B_E]$ must be defined for each time step using a contact algorithm. The first equation in (11.5) can be recast to obtain

$$\{\ddot{u}\}_n = [M]^{-1}\left[\{f\}_n - [C]\{\dot{u}\}_n - [K]\{u\}_n - [B_E]^T_{n+1}\{\lambda\}_n\right]. \tag{11.6}$$

Hence, by considering the Newmark central difference scheme for the explicit integration

$$\{\ddot{u}\}_n = \frac{1}{h^2}\left[\{u\}_{n+1} - 2\{u\}_n + \{u\}_{n-1}\right], \tag{11.7}$$

by substituting Eq. (11.7) in Eq. (11.6) and by using $\{\dot{u}\}_n = \left(\{u\}_n - \{u\}_{n-1}\right)/h$ one obtains

$$\{u\}_{n+1} = h^2[M]^{-1}\left[\{f\}_n - \frac{1}{h}[C]\left(\{u\}_n - \{u\}_{n-1}\right) - [K]\{u\}_n\right] + 2\{u\}_n - \{u\}_{n-1} - h^2[M]^{-1}[B_E]^T_{n+1}\{\lambda\}_n. \tag{11.8}$$

The displacement $\{u\}$ at time $n+1$ can be rewritten as the sum of a predictor $\{u\}^\star_{n+1}$ and a corrector $\{u\}^c_{n+1}$:

$$\{u\}_{n+1} = \{u\}^\star_{n+1} + \{u\}^c_{n+1} \tag{11.9}$$

where

$$\{u\}^\star_{n+1} = h^2[M]^{-1}\left[\{f\}_n - \frac{1}{h}[C]\left(\{u\}_n - \{u\}_{n-1}\right) - [K]\{u\}_n\right] + 2\{u\}_n - \{u\}_{n-1} \tag{11.10}$$

and

$$\{u\}^c_{n+1} = -h^2[M]^{-1}[B_E]^T_{n+1}\{\lambda\}_n. \tag{11.11}$$

The quantity in Eq. (11.10) provides a prediction of displacement that does not account for contact forces. It must be corrected by the quantity in Eq.(11.11) that represents the incremental displacement associated with the contact forces. To evaluate the corrector in Eq. (11.11), contact forces must be estimated and the time varying constraints must be defined.

The contact forces intensities $\{\lambda\}_n$ can be obtained by substituting Eq. (11.9) in the compatibility condition of Eq. (11.5) and by considering the expression of the corrector in Eq. (11.11):

$$\{\lambda\}_n = \left(h^2[B_C]_{n+1}[M]^{-1}[B_E]^T_{n+1}\right)^{-1}[B_C]_{n+1}\{u\}^\star_{n+1}. \tag{11.12}$$

Note that $\{\lambda\}_n$ depends on matrices $[B_C]$ and $[B_E]$ at time t_{n+1}. They can be evaluated by using different approaches:

- A simplified contact assumption, introduced in [16], based on the rigid, a priori defined, relative motion of component subsystems due to the kinematic boundary conditions ("Subs1").

- A more realistic contact assumption that considers the deformation of the contacting bodies but neglects the variation of the normal force direction due to the deformation ("Subs2").
- A furtherly refined contact algorithm, introduced in [19] that accounts for all the implications of body deformations ("Subs3").

All the proposed approaches are based on the assumption that contacting bodies are in sliding contact (neither sticking nor detachment is possible), and this assumption must be verified during the simulation. In the following section, the results obtained by these substructuring based procedures are compared with the truth solution ("FiLm") provided by the "Forward Increment Lagrange Multiplier Method" [32], a well assessed numerical method for dynamic contact problems that would manage any possible contact status (sliding, sticking, or detachment).

11.3 Results

The mechanical system considered in this work is identical to the model used in [18], and it is composed of two substructures connected through a time-variant interface. As shown in Fig. 11.1, the inclined beam 1 represents the first substructure, it is made of Aluminum Alloy and is subjected to time-variant boundary conditions, vertical force F_y and horizontal velocity v_y, acting on point A; the horizontal cantilever beam 2 represents the second substructure, and it is made of Polycarbonate. The tip C of beam 1, initially distant s_0 from the fixed end B of beam 2, slides along the upper edge of beam 2 and a constant friction coefficient μ is considered at the interface. Geometric dimensions are reported in Table 11.1 while material properties and boundary conditions are summarized in Table 11.2. Note that a viscous proportional damping with coefficients α and β is considered and that, in order to limit the amplitude of transient vibrations, generated by the application of the boundary conditions, the vertical load F_y, the velocity v_x, and the friction coefficient μ are smoothly applied and higher damping coefficients ($\alpha_0 = 40\,\mathrm{s}^{-1}$ and $\beta_0 = 4.0e - 8\,\mathrm{s}$) are adopted during the variation of forces and velocity.

The two beams are modeled by plane stress elements in ANSYS, and a modal reduction of both substructure is performed using the Craig Bampton method with 20 modes for each substructure. Therefore, beside the modal degrees of freedom, only the physical nodes on which the boundary conditions apply and the ones involved in the contact are maintained for the explicit integration, for a global size of the model of 243 DoFs. As shown in [18], when the oblique beam slides along the horizontal one, this system presents possible dynamic instabilities due to mode coupling at some position of the contact point. The stability of the system can be evaluated by coupling the two substructures for every position using the primal assembly, as proposed in [16] and by computing the corresponding complex eigenvalues. It is expected that, when the system enters into an instability region, the vibration amplitude increases exponentially and that it extinguishes when the instability region is overcome.

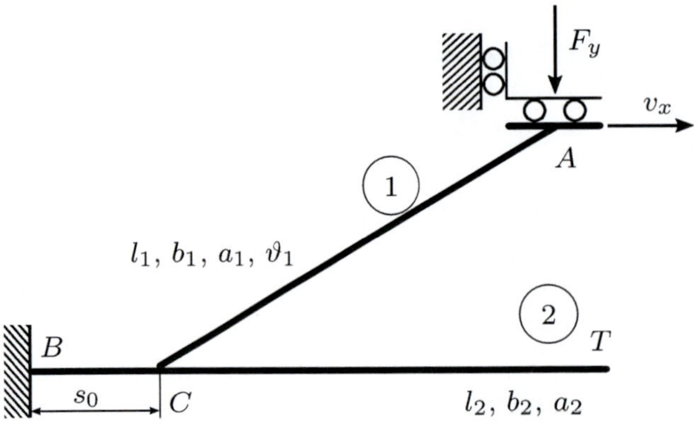

Fig. 11.1 Beam on beam system

Table 11.1 Geometrical characteristics of the two beams

Dimensions		Beam 1	Beam 2	
Length	l	75	100	mm
Thickness	b	1.5	2	mm
Width	a	10	40	mm
Angle	ϑ	30	0	degree

Table 11.2 Material properties and boundary conditions

Dimensions		Beam 1	Beam 2	
Young mod	E	71	2.38	GPa
Poisson ratio	ν	0.33	0.40	
Density	ρ	2770	1200	kg/m^3
$[M]$ prop. damping	α	0.2	40	s^{-1}
$[K]$ prop. damping	β	$4.0e-8$	$4.0e-8$	s
Friction coefficient	μ	0.30		
Force	F_y	−0.20		N
Velocity	v_x	30		mm/s

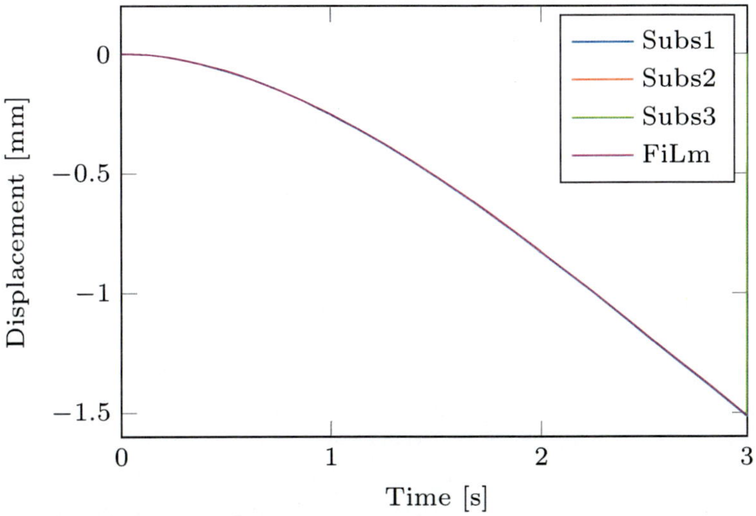

Fig. 11.2 Vertical displacement of point T obtained using the three substructuring based method compared with the truth solution

Figure 11.2 shows the comparison of the vertical displacement of the free end of the horizontal beam for the three substructuring based procedure, with respect to the truth solution. Displacement results show a progressive lowering of the point T due to the displacement of the vertical force exchanged between the two beams. This result confirms that if the vibrations are negligible with respect to the quantity of interest, the three substructuring based methods, built on three different contact assumption, provide the same information.

The four plots in Fig. 11.3 show a comparison of the vertical acceleration of the free end of the horizontal beam obtained using the three substructuring based procedure with the truth acceleration response provided by the FiLm method. The truth solution in Fig. 11.3d shows the typical bursts of vibration associated with the crossing of instability region. The acceleration response obtained with the Subs1 method is shown in Fig. 11.3a, and it highlights that this method does not capture the instability burst of vibrations. Therefore, although the Subs1 method, as it is formulated in [16], is capable of detecting the effects of time-variant system configuration due to the macroscopic sliding between substructures, and their effects in the global dynamics [18], it does not provide reliable results in case of contact dynamic instabilities. Conversely, the responses obtained with the methods Subs2 and Subs3, shown in Fig. 11.3b, c respectively, highlight that accounting for system deformation in the contact assumption is a crucial aspect for a correct estimation of energy exchanged among substructures [25] and a consequent correct detection of friction induced instabilities. In particular, the method Subs2, that does not account for the effects of the system deformation on the normal force direction, provides a non-correct estimation of the maximum amplitude reached by the self-excited vibration bursts (they are over-estimated in the presented case), while the method Subs3 is able to capture both the occurrence and the intensity of the unstable friction induced vibrations.

11.4 Conclusions

In this paper a substructuring based method, previously proposed by the authors to deal with time-variant systems composed of time-invariant substructures subjected to time-variant coupling conditions, is used to evaluate the response of a mechanical system with a sliding contact interface. The effects of different assumptions made in the formulation of the time-variant

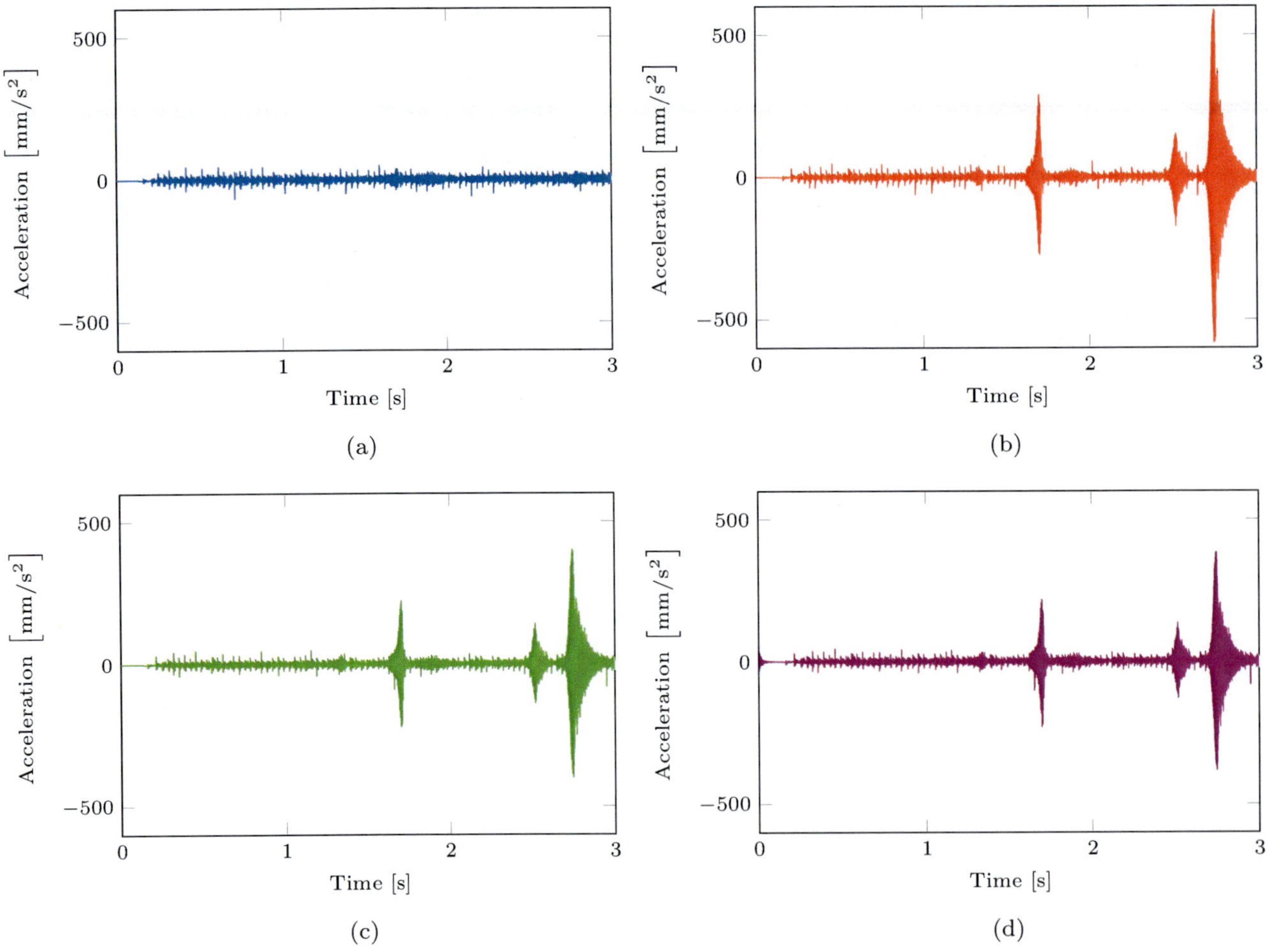

Fig. 11.3 Vertical acceleration of point T. (**a**) Subs1 method. (**b**) Subs2 method. (**c**) Subs3 method. (**d**) FiLm method

compatibility and equilibrium conditions are investigated, by comparing the results provided by the substructuring based methods with the truth response of the system. The results show that depending on the complexity of the phenomena under investigation different assumption can be made in the substructuring approach. In particular, by assuming that, the time-variant compatibility and equilibrium conditions, arising from sliding contact, are a priori known, a correct estimation of the variation of the global dynamics can be obtained. This approach is suitable for the analysis of configuration dependent dynamics of a time-variant system. However, a reliable analysis of more complex phenomena, related to time-variant system with sliding friction interface, in the framework of dynamic substructuring, requires a more refined contact algorithm. The results presented highlight that, to detect and correctly retrace unstable friction induced vibrations, the time-variant coupling conditions must be defined accounting also for the deformation of coupled bodies.

Acknowledgments This research is supported by University of Rome La Sapienza and University of L'Aquila.

References

1. de Klerk, D., Rixen, D.J., Voormeeren, S.: General framework for dynamic substructuring: history, review, and classification of techniques. AIAA J. **46**(5), 1169–1181 (2008)
2. Sjövall, P., Abrahamsson, T., Substructure system identification from coupled system test data. Mech. Syst. Signal Proces. **22**(1), 15–33 (2008)
3. D'Ambrogio, W., Fregolent, A.: Inverse dynamic substructuring using direct hybrid assembly in the frequency domain. Mech. Syst. Signal Proces. **45**(2), 360–377 (2014)
4. Mayes, R.L., Ross, M.R.: Advancements in hybrid dynamic models combining experimental and finite element substructures. Mech. Syst. Signal Proces. **31**, 56–66 (2012)
5. Rixen, D.: A dual Craig-Bampton method for dynamic substructuring. J. Comput. Appl. Math. **168**(1–2), 383–391 (2004)

6. D'Ambrogio, W., Fregolent, A.: Are rotational DoFs essential in substructure decoupling?. In: Allen, M., Mayes, R. Rixen, D. (eds.) Dynamics of Coupled Structures, vol. 1. Conference Proceedings of the Society for Experimental Mechanics Series, pp. 27–36. Springer International Publishing, Cham (2014)
7. Kuether, R.J. Allen, M.S.: Nonlinear modal substructuring of systems with geometric nonlinearities. In:s 54th AIAA/ASME/ASCE/AHS/ASC Structures, Structural Dynamics, and Materials Conference, p. 1521 (2013)
8. Kalaycıoğlu, T., Özgüven, H.N.: Nonlinear structural modification and nonlinear coupling. Mech. Syst. Signal Proces. **46**(2), 289–306 (2014)
9. Latini, F., Brunetti, J., D'Ambrogio, W. Fregolent, A.: Substructures' coupling with nonlinear connecting elements. Nonlinear Dyn. **99**(2), 1643–1658 (2020)
10. Krattiger, D., Wu, L., Zacharczuk, M., Buck, M., Kuether, R., Allen, M., Tiso, P. Brake, M.: Interface reduction for Hurty/Craig-Bampton substructured models: Review and improvements. Mech. Syst. Signal Proces. **114**, 579–603 (2019)
11. Brunetti, J., D'Ambrogio, W., Fregolent, A. Latini, F.: Substructuring using NNMS of nonlinear connecting elements. Lecture Notes in Mechanical Engineering, pp. 1426–1440 (2020)
12. Latini, F., Brunetti, J., Kwarta, M., Allen, M.S., D'Ambrogio, W., Fregolent, A.: Experimental results of nonlinear structure coupled through nonlinear connecting elements. In: Proceedings of ISMA 2020 – International Conference on Noise and Vibration Engineering and USD 2020 – International Conference on Uncertainty in Structural Dynamics (2020)
13. Rixen, D.J. van der Valk, P.L.: An Impulseb based substructuring approach for impact analysis and load case simulations. J. Sound Vib. **332**(26), 7174 – 7190 (2013)
14. van der Valk, P.L., Rixen, D.J.: An impulse based substructuring method for coupling impulse response functions and finite element models. Comput. Methods Appl. Mech. Eng. **275**(Supplement C), 113–137 (2014)
15. Carassale, L., Silvestri, P., Lengu, R. Mazzaron, P.: Modeling rail-vehicle coupled dynamics by a time-varying substructuring scheme. In: Linderholt, A., Allen, M.S., Mayes, R.L. Rixen, D. (eds.) Dynamic Substructures, vol. 4, pp. 167–171. Springer International Publishing, Cham (2020)
16. Brunetti, J., D'Ambrogio, W. Fregolent, A.: Dynamic coupling of substructures with sliding friction interfaces. Mech. Syst. Signal Proces. **141**, 106731 (2020)
17. Brunetti, J., D'Ambrogio, W., Fregolent, A.: Contact problems in the framework of dynamic substructuring. In: Proceedings of ISMA 2018 – International Conference on Noise and Vibration Engineering and USD 2018 – International Conference on Uncertainty in Structural Dynamics, pp. 3987–3998 (2018)
18. Brunetti, J., D'Ambrogio, W., Fregolent, A.: Dynamic substructuring with time variant coupling conditions for the analysis of friction induced vibrations. In: Proceedings of ISMA 2020 – International Conference on Noise and Vibration Engineering and USD 2020 – International Conference on Uncertainty in Structural Dynamics, pp. 3023–3032 (2020)
19. Brunetti, J., D'Ambrogio, W., Fregolent, A.; Friction-induced vibrations in the framework of dynamic substructuring. Nonlinear Dynamics, **103**(4), 3301–3314 (2021). https://doi.org/10.1007/s11071-020-06081-9
20. Akay, A.: Acoustics of friction. J. Acoust. Soc. Am. **111**(4), 1525–1548 (2002)
21. Renouf, M., Massi, F., Saulot, A., Fillot, N.: Numerical tribology of dry contact. Tribol. Int. **44**(7–8), 834–844 (2011)
22. Batailly, A., Legrand, M., Cartraud, P., Pierre, C.: Assessment of reduced models for the detection of modal interaction through rotor stator contacts. J. Sound Vib. **329**(26), 5546–5562 (2010)
23. Ibrahim, R.A.: Friction-induced vibration, chatter, squeal, and chaos – part I: mechanics of contact and friction. Appl. Mech. Rev. **47**(7), 209–226 (1994)
24. Tonazzi, D., Passafiume, M., Papangelo, A., Hoffmann, N., Massi, F.: Numerical and experimental analysis of the bi-stable state for frictional continuous system. Nonlinear Dyn. **102**(3), 1361–1374 (2020)
25. Brunetti, J., Massi, F., D'Ambrogio, W. Baillet, L.: Steady state of modal coupling instabilities as a dynamic energy equilibrium. In: Proceedings of ISMA 2014 – International Conference on Noise and Vibration Engineering, Leuven, pp. 1827–1842 (2014)
26. Adams, G.G.: Steady sliding of two elastic half-spaces with friction reduction due to interface stick-slip. J. Appl. Mech. **65**(2), 470–475 (1998)
27. Tonazzi, D., Massi, F., Culla, A., Baillet, L., Fregolent, A, Berthier, Y.: Instability scenarios between elastic media under frictional contact. Mech. Syst. Signal Proces. **40**(2), 754–766 (2013)
28. Papangelo, A., Hoffmann, N., Grolet, A., Stender, M., Ciavarella, M.: Multiple spatially localized dynamical states in friction-excited oscillator chains. J. Sound Vib. **417**, 56–64 (2018)
29. Sinou, J.-J., Thouverez, F., Jézéquel, L.: Analysis of friction and instability by the centre manifold theory for a non-linear sprag-slip model. J. Sound Vib. **265**(3), 527–559 (2003)
30. Hoffmann, N., Gaul, L.: A sufficient criterion for the onset of sprag-slip oscillations. Arch. Appl. Mech. **73**(9–10), 650–660 (2004)
31. Di Bartolomeo, M., Lacerra, G., Baillet, L., Chatelet, E., Massi, F.: Parametrical experimental and numerical analysis on friction-induced vibrations by a simple frictional system. Tribol. Int. **112**, 47–57 (2017)
32. Carpenter, N.J., Taylor, R.L., Katona, M.G.: Lagrange constraints for transient finite element surface contact. Int. J. Numer. Methods Eng. **32**(1), 103–128 (1991)

Chapter 12
Road Noise: Embedding Suspension Test Benches in Sound & Vibration Design using virtual points and the Transfer Path Analysis Framework

Maarten V. van der Seijs, Julie M. Harvie, and David P. Song

Abstract Many Car Manufacturers (OEMs) use dedicated test bench setups for testing and developing components separately from their vehicle platforms. Relating these results to full-vehicle assemblies is not immediately possible because of the difference in boundary conditions and incompatibility between test bench and vehicle measurement points.

This study focuses on the use of tire noise test benches in the engineering of full-vehicle NVH. We show how virtual points, easily obtained using DIRAC, ensure compatibility between various test assemblies. A total of 10 contact points between the suspension parts and the bodywork is described using 3, 5 or 6 degrees of freedom per point. The TPA framework is extensively used to create and compare source descriptions: direct blocked forces from a rigid test bench, in situ blocked forces from a compliant test bench, and several component TPA approaches on the full vehicle. Blocked forces are estimated both before and after the bushings. Results are evaluated for simulation of road noise at the driver's ear in the vehicle.

Keywords NVH · Blocked forces · Transfer path analysis · Virtual point · Test bench

12.1 Introduction

Road noise has known to be one the major noise sources in a modern car. While drivetrain noise levels nowadays go down considerably – already due to replacement of the combustion engine by an electric one – noise induced by the tire-road contact remains an important Noise, Vibration and Harshness (NVH) aspect. For a typical consumer car, the structure-borne noise contribution sits predominantly in the frequency range from 20 to 500 Hz, whereafter the airborne contributions gradually take over. Within this bandwidth, one sometimes speaks of "booming" noise between approximately 20–200 Hz, a smaller band associated with the tire cavity typically around 200–250 Hz and "rumble" noise up to 500 Hz. Mitigating road noise means reducing the vibration propagation from the tire-road contact to the driver's and passengers' ears. A major point of focus for an OEM is the suspension subsystem, which comprises the knuckle-wheelhub assembly, the spring/damper, and the various linkages that define the suspension mechanism. Figure 12.1 shows a schematical depiction of a typical rear suspension system.

Many of the suspension linkages are interconnected by rubber bushings. Each of those bushings is designed carefully for optimal performance, considering different aspects such as handling, drive comfort, and of course NVH performance. In this design, one often finds trade-offs between the mentioned aspects, which limit the design space to certain dynamic stiffness ranges in the axial and radial directions of the bushing. As usual in dynamics, changes to particular bushings affect the transmission of road noise more than others. The optimization of the suspension for NVH performance is therefore a multi-variable challenge and needs a process that is able to evaluate changes on the full-vehicle level.

In this chapter, we describe a process that uses measurements on a rear-wheel suspension test bench, vehicle noise transfer functions, and a state-of-art approach using virtual points [1] to ensure compatibility of the two setups. Essentially, we demonstrate application of component-based TPA [2], using several approaches to estimate the blocked forces.

M. V. van der Seijs (✉) · J. M. Harvie
VIBES Technology, Delft, The Netherlands
e-mail: mvanderseijs@vibestechnology.com

D. P. Song
Hyundai Motor Company – NVH Research Lab, Hwaseong-si, Gyeonggi-do, South Korea

M. S. Allen et al. (eds.), *Dynamic Substructures, Volume 4*, Conference Proceedings of the Society for Experimental Mechanics Series, https://doi.org/10.1007/978-3-030-75910-0_12

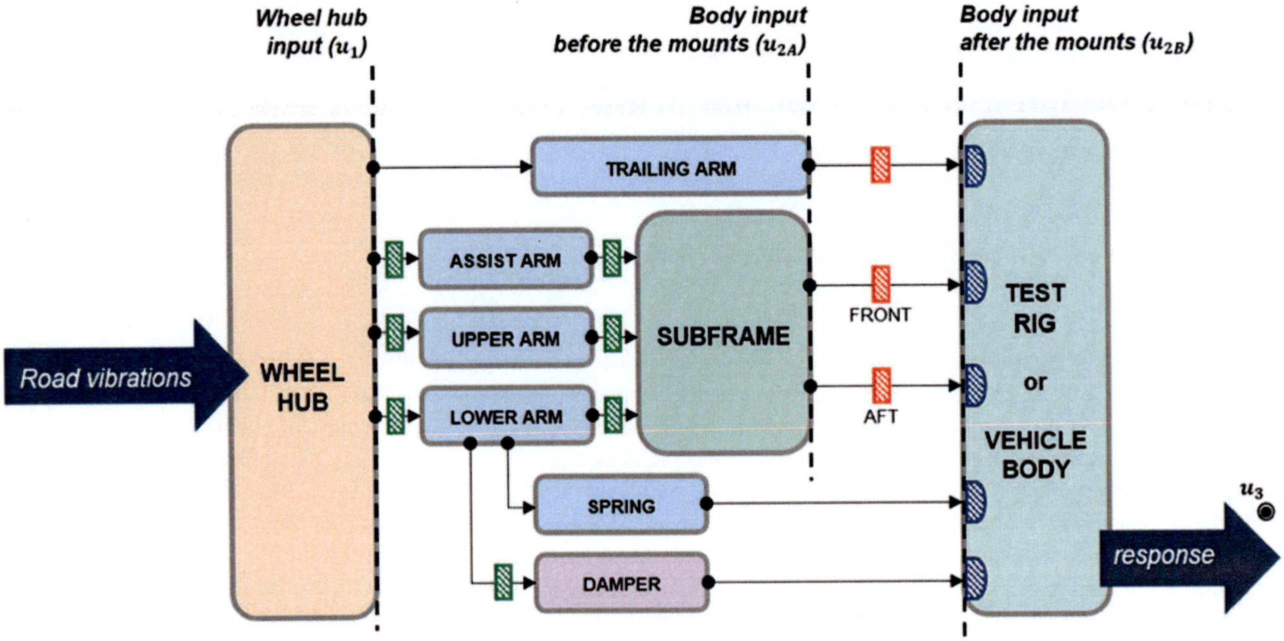

Fig. 12.1 Schematic overview of the rear-wheel supension system for a single side

12.2 Approach

The component TPA approach with use of virtual points allows for a unified TPA simulation procedure. The blocked forces f_2^{bl} will be applied to the full-vehicle transfer functions Y_{32}^{AB} to predict the driver's ear responses u_3 in the vehicle:

$$u_3 = Y_{32}^{\mathrm{AB}} f_2^{\mathrm{bl}} \tag{12.1}$$

Because of the presence of rubber bushings, it is relevant to consider the point of application of blocked forces in terms of before and after the mounts:

1. *After the mounts* (f_{2B}^{bl}): blocked forces include the effects of the bushings to the bodywork. The forces can be applied straight away to the full vehicle bodywork on the passive side.
2. *Before the mounts* ($f_{2A}^{\mathbf{bl}}$): blocked forces do not include the effect of the bushings. This allows changing the bushings by virtually adjusting the Frequency Response Functions (FRF) of the full vehicle and applying the blocked forces to this FRF on the active side.

In this study, both variants are considered. The configuration using passive-side blocked forces after the mounts is used as the base case, as it aligns with the rigid rig configuration.

12.2.1 Connection Points

The compatibility of test rig forces and vehicle transfer functions is critical to successful application of Equation 12.1. In this study, we used virtual points at prescribed locations to have a common interface between all test assemblies. Figure 12.2 shows the five connection points of the rear suspension to the vehicle bodywork: the trailing arm, subframe front and rear mounts, the strut with damper, and the spring. In total, we consider the 10 coupling points for the left and right sides simultaneously.

As indicated, the modeling strategy can be chosen for each coupling point differently. The four subframe mounts typically only exhibit translational motion and can be well described by 3-DoF translational dynamics. For the trailing arm and the strut, additional rotational DoF have been added for optional moment information. This was also done to match the instrumentation of the force sensors on the test bench: for these points, two sensors per coupling point were used which allow estimation of two additional moments.

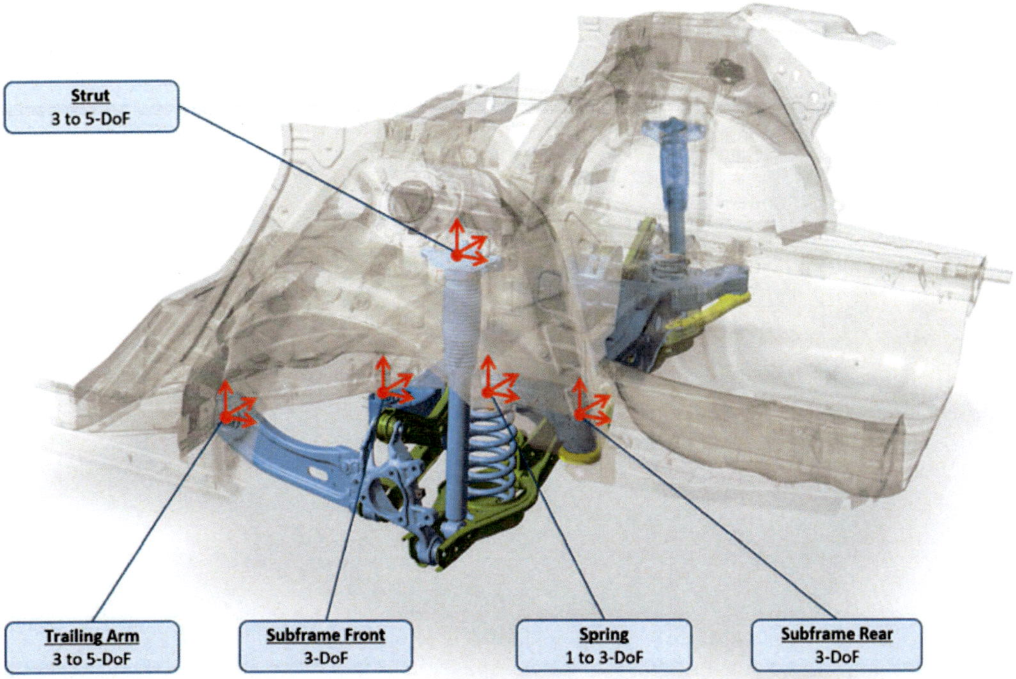

Fig. 12.2 The five connection points of the rear suspension and their modeling strategy using virtual points

Fig. 12.3 The rigid test rig. Left: the tire on the dyno. Center: two force gauges installed at the trailing arm passive-side connection point. Right: the same setup in the DIRAC modeling environment showing the location of the force gauges

12.2.2 Rigid Rig

The rigid test rig has been used and described in previous studies [3, 4]. The rig is optimized for direct force measurement and has force gauges installed at the connecting points to the body *after* the bushings; see Fig. 12.3. The measured forces can thus be interpreted as direct blocked forces *after the mount*, or f_{2B}^{bl} in notation of the TPA Framework [2]. Although this is theoretically correct, successful application warrants consideration of some important practical aspects as discussed below.

Processing of the Force Gauge Data The force gauges are installed in the most practical locations, namely between the rigid rig adapters and the bushings and along the direction of the bolts that normally connect the suspension to the vehicle. This means that every coupling point has a somewhat local coordinate frame that may be cumbersome to match with any transfer function measured on the vehicle. The virtual point approach is therefore used to transform all measured operational forces into the global coordinate frame, as well as to sum them in a kinematically correct manner. In this case of measured forces, we use the transpose of the Interface Displacement Mode (IDM) matrix for the force gauges g_{2B} to the virtual points [5]:

$$\mathrm{f}_{2B}^{bl} = (\mathrm{R_f})^T g_{2B}^{oper} \tag{12.2}$$

The IDM matrix is automatically computed using DIRAC for the design of experiment, as shown in Fig. 12.3. Equation 12.2 performs the kinematically correct summation of all measured force channels to create the virtual point blocked forces and moments. For example, this also reduces the total of 6 DoF of the trailing arm gauges to 3–5 virtual point forces in the global reference frame.

Rigidness of the Test Rig Direct blocked-force measurement is known to be most effective for low frequencies. The validity of the rigid boundary assumption breaks down with increasing frequencies, due to flexibility of the test rig. As shown in Fig. 12.3, the rigid rig adapters are optimized for maximum stiffness, being constructed from solid steel. Nevertheless, the first eigenmodes of the test rig may have a degrading effect on the blocked forces.

12.2.3 Soft Rig

One of the focus points of the study was to also investigate an approach using more flexible test rig adapters. The resulting "soft rig" construction is shown in Fig. 12.4. In this case, accelerometers at the passive side of the connection points are used to calculate blocked forces at the passive side in a matrix-inverse way, a method known as in situ characterization [6]. A great benefit is now that both blocked forces before and after the mount can be computed with the same ease. Both cases use the same passive-side indicator responses; the difference lies in the virtual point forces (2A or 2B) of the FRF matrix to invert:

$$\text{Active} - \text{side} : f_{2A}^{bl} = \left(Y_{42A}^{AR}\right)^{+} u_4 \tag{12.3}$$

$$\text{Passive} - \text{side} : f_{2B}^{bl} = \left(Y_{42B}^{AR}\right)^{+} u_4 \tag{12.4}$$

The FRF matrices with virtual point force columns have been created using DIRAC. These are sometimes called "right-sided" virtual point FRF matrices, as the virtual point transformation is only applied to the FRF columns, which is sufficient for in situ calculation [2]. The overdetermination of the matrix inversion is kept at 150–200%.

12.2.4 Vehicle

FRF and operational tests have also been conducted on the full vehicle. Full-vehicle FRFs were obtained similar as for the compliant test rig, namely for impact points before and after the mounts to the bodywork and again transformed to virtual points. These virtual point forces are now directly compatible with the virtual point blocked forces coming from the test rigs, as they are both defined in the unique coordinate frame of the virtual points.

The FRF tests have again been conducted using DIRAC, connected to a Müller-BBM PAK MK2 system for acquisition of about 200 channels, including some microphones at the ear positions. In order to excite the full bandwidth of 0–500 Hz, two impact hammers have been used with different qualities: a heavier PCB 086D05 hammer with rubber tip to cover the low frequencies (0–200 Hz) with sufficient energy and a lighter PCB 086C03 hammer with nylon tip to excite up to several kHz. DIRAC processes the data correspondingly using low- and high-pass filtering and makes sure that the measurement data is correctly added in the estimation of the transfer functions. The result is near-perfect coherence and signal-to-noise for all FRF functions in the frequency range of interest. This is illustrated in Fig. 12.5.

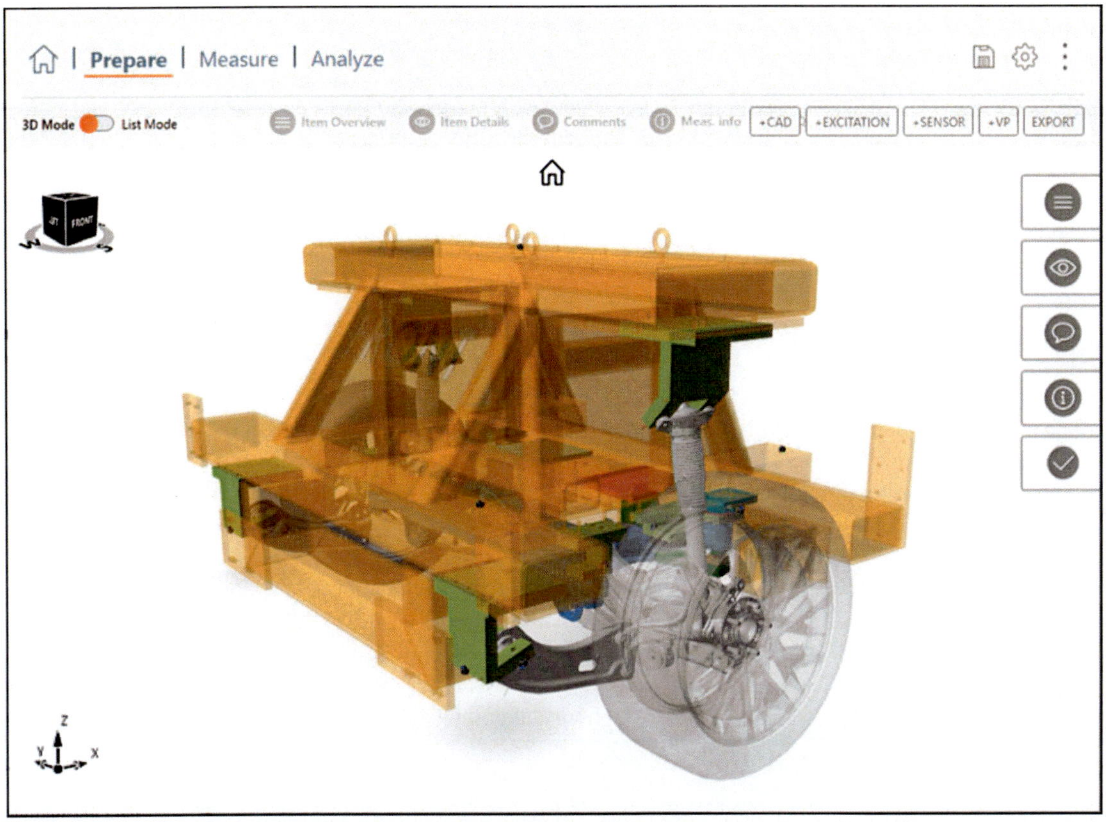

Fig. 12.4 The soft rig setup in DIRAC. The flexible adapters are shown in green

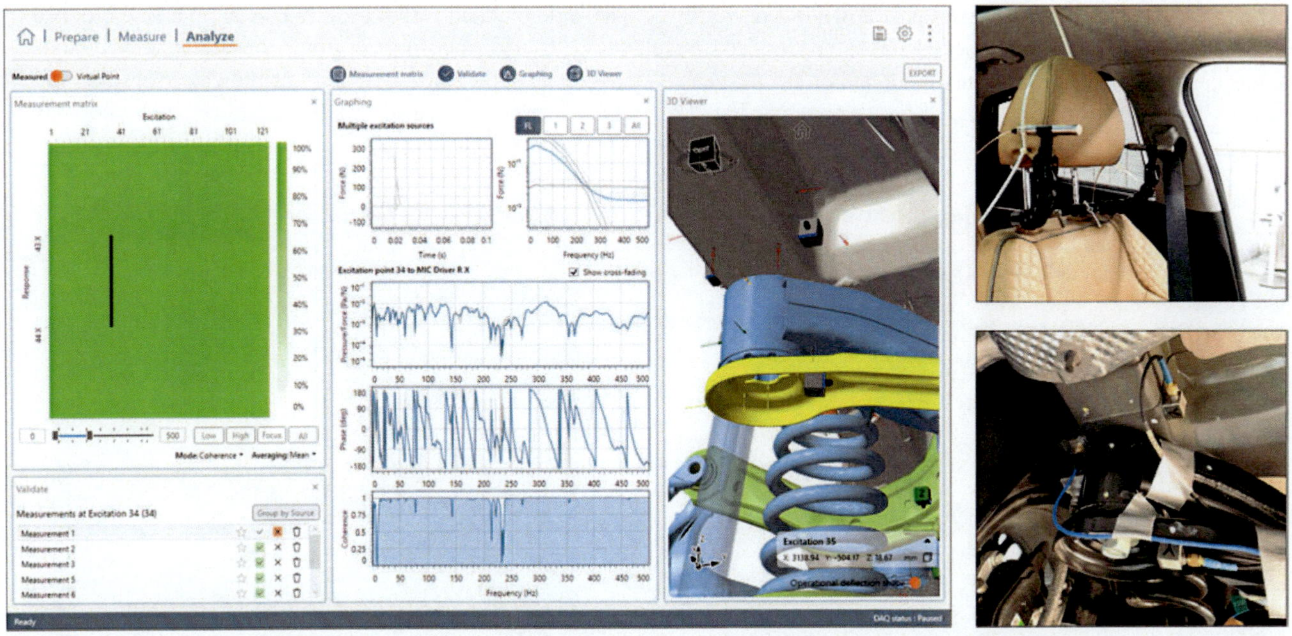

Fig. 12.5 Full-vehicle FRF measurement in DIRAC. Two impact hammers were used to properly excite the full frequency range of 0-500Hz; their data (gray curves) is merged directly in the estimation of the transfer functions (blue curve)

12.3 Results

Let us now discuss the application of the methods described above. The operational cycle used for all test assemblies is a runup of the rear wheels on a dyno from 20 to 120 km/h. The recorded time data is converted to 1-second blocks with 50% overlap and Hann-windowed before being computed into frequency spectra. Whereas DIRAC is used for acquisition and virtual point transformation of all FRFs, all processing and TPA simulation are performed using the VIBES Toolbox for MATLAB.

12.3.1 Rigid Rig: Direct Blocked Forces

Let us first look at the application of the direct blocked forces from the rigid test rig. In this case, the direct forces are obtained after the mounts, so they are multiplied with the vehicle FRFs on the passive side. Figure 12.6 shows the simulation at the left driver ear compared to the direct measurement as a Campbell diagram, at the same color scale. It is clear to see that all relevant noise phenomena in the car are well represented by the test rig blocked forces applied to the vehicle noise transfer functions.

Another way of comparing the component TPA prediction with the measurement is by looking at the time-averaged spectra of driver-ear sound pressure levels in dB(A) over the full runup from 20 to 120 km/h. The result of the rigid rig direct blocked forces is shown in Fig. 12.7a in red, where comparison is made with the vehicle measurement for a similar runup (black). The horizontal grid lines represent intervals of 10 dB(A). There is clearly a very high agreement, which demonstrates the power and simplicity of direct blocked-force TPA through virtual points.

12.3.2 Soft Rig: In Situ Blocked Forces

For the soft rig application, two cases are shown in Fig. 12.7b: passive-side blocked forces (yellow) and active-side blocked forces (red). The passive-side blocked forces are indeed a result of VP-transformed impact points on the adapter blocks, whereas the active-side impacts are all applied on the suspension parts close to the mounts. The same indicator sensors were used for both calculations, namely all on the passive side. The matrix-inverse calculation has been calibrated by choosing

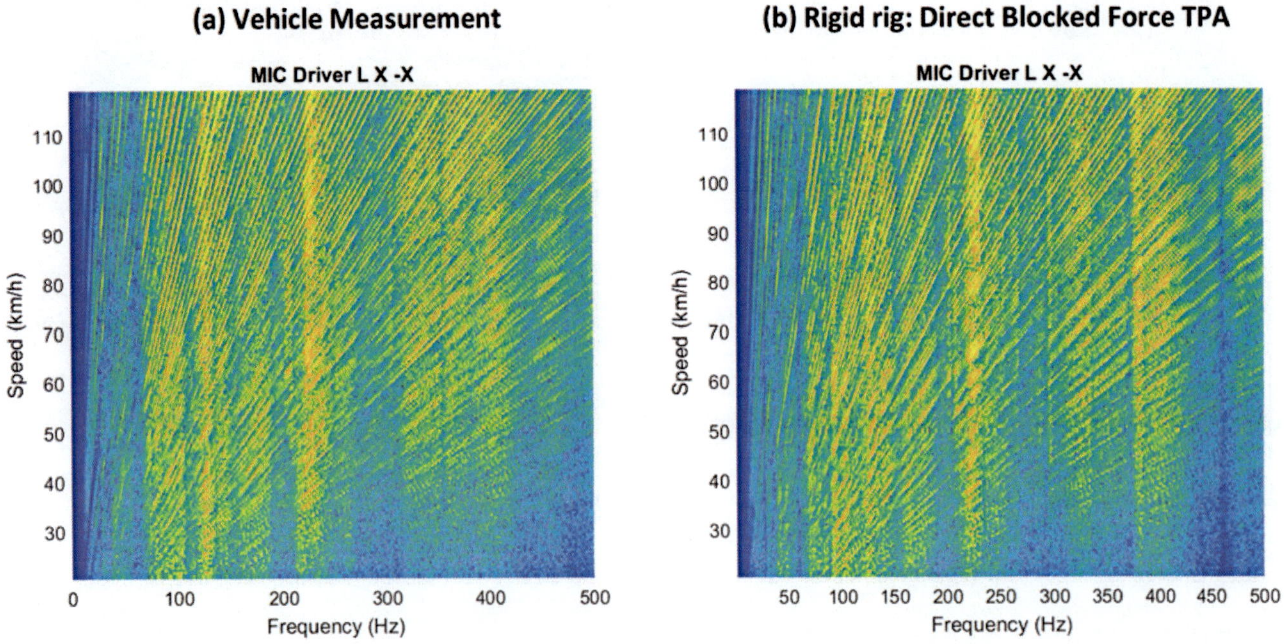

Fig. 12.6 Validation measurement (**a**) compared to the prediction (**b**) of the direct blocked forces from the rigid rig

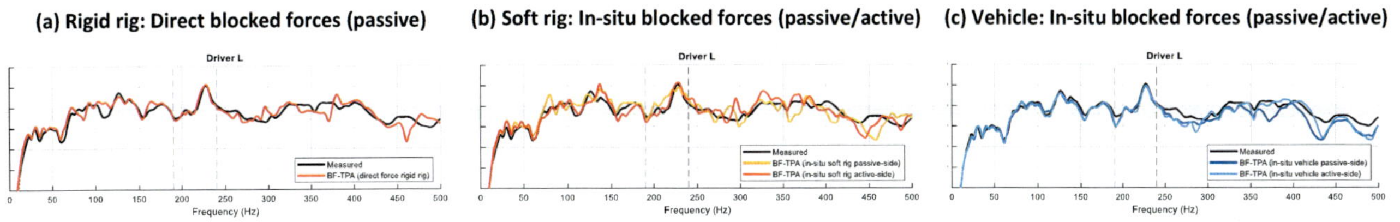

Fig. 12.7 Result of blocked force application as dB(A) spectra. (**a**) Rigid rig direct blocked forces. (**b**) Soft rig in situ passive (yellow) and active (red) sides. (**c**) Vehicle in situ passive (blue) and passive (cyan) sides

the amount of singular values based on the number of active DoF in the inverse problem, using independent control points for validation.

Both predictions from the soft rig in situ blocked forces show a good match with the general trend of the vehicle measurement. Compared with the direct force application, the curves tend to wobble a bit more. This is most likely explained by the sensitivity to anti-resonances in the soft-rig FRFs, which plays a major role as it is inverted in the blocked-force calculation. A benefit over the direct blocked-force approach is the availability of blocked forces before the bushings. These can be used in conjunction with vehicle FRFs where now the bushing stiffness can be altered by dynamic substructuring, leading to physically correct simulation of bushing changes.

12.3.3 Vehicle: In Situ Blocked Forces

Finally, we present a similar in situ approach for blocked forces calculated from the vehicle runup. This can be seen as the true *in situ* approach: the full vehicle on the dyno is the only test assembly under consideration. The prediction is shown in Fig. 12.7c for passive-side blocked forces (blue) and active-side blocked forces (cyan). The match up to 250 Hz is close to perfect for both cases; at higher frequencies the agreement is also good. The active- and passive-side blocked forces show similar performance.

12.4 Conclusions

Several methods for blocked-force characterization and component-based TPA have been discussed for the application of road noise NVH. A workflow has been described that uses virtual points to establish compatibility between two test rig setups for the suspension system and full vehicle. It has been demonstrated that test rig measurements can indeed be used for accurate prediction of NVH levels in the vehicle. The direct blocked-force method has shown to produce good results, proving the rigidness of the test rig to be sufficient in the frequency range of interest. The in situ methods using acceleration measurement and matrix inversion allow forces to be determined before or after the mount, both from test rig and vehicle measurements.

Acknowledgments We like to acknowledge NVH Research Lab of Hyundai Motor Company for funding this research. We like to thank Müller-BBM VibroAkustik Systeme Korea for their support in the measurement campaigns and coordination.

References

1. van der Seijs, M.V.: Experimental dynamic substructuring: analysis and design strategies for vehicle development, PhD Dissertation. Delft University of Technology (2016)
2. van der Seijs, M.V., Rixen, D.J.E., De Klerk, D.: General framework for transfer path analysis: History, theory and classification of techniques. Mech. Syst. Signal Process. **68**, 217–244 (2016)
3. Kang, Y.J., Song, D.P., Ih, K.D.: Estimation of body input force transmission change due to parts' modification using the impedance method under rolling excitation. Proc. Institut. Mech. Eng. Part D: J. Automob. Eng. **233**(2), 363–377 (2019)
4. Song, D.P., Min, D., Kang, Y.J., Cho, M., Kim, H.G., Ih, K.D.: A methodolgy for evaluating the structure-borne road noise prior to a prototype vehicle using direct force measured on a suspension rig. Noise Cont. Eng. J. **64**(3), 295–304 (2016)

5. van der Seijs, M.V., van den Bosch, D.D., Rixen, D.J., de Klerk, D.: An improved methodology for the virtual point transformation of measured frequency response functions in dynamic substructuring. In: COMPDYN 2013: 4th ECCOMAS thematic conference on computational methods in structural dynamics and earthquake engineering. Kos, Greece (2013)
6. Moorhouse, A.T., Elliott, A.S., Evans, T.A.: In situ measurement of the blocked force of structure-borne sound. J. Sound Vib. **325**(4-5), 679–685 (2009)